Practical R for Mass Communication and Journalism

Chapman & Hall/CRC
The R Series

Series Editors

Recently Published Titles

Basics of Matrix Algebra for Statistics with R
Nick Fieller

Introductory Fisheries Analyses with R
Derek H. Ogle

Statistics in Toxicology Using R
Ludwig A. Hothorn

Spatial Microsimulation with R
Robin Lovelace, Morgane Dumont

Extending R
John M. Chambers

Using the R Commander: A Point-and-Click Interface for R
John Fox

Computational Actuarial Science with R
Arthur Charpentier

bookdown: Authoring Books and Technical Documents with R Markdown,

Yihui Xie

Testing R Code

Richard Cotton

R Primer, Second Edition
Claus Thorn Ekstrøm

Flexible Regression and Smoothing: Using GAMLSS in R
Mikis D. Stasinopoulos, Robert A. Rigby, Gillian Z. Heller, Vlasios Voudouris, and Fernanda De Bastiani

The Essentials of Data Science: Knowledge Discovery Using R
Graham J. Williams

blogdown: Creating Websites with R Markdown
Yihui Xie, Alison Presmanes Hill, Amber Thomas

Handbook of Educational Measurement and Psychometrics Using R

Christopher D. Desjardins, Okan Bulut

Displaying Time Series, Spatial, and Space-Time Data with R, Second Edition

Oscar Perpinan Lamigueiro

Reproducible Finance with R
Jonathan K. Regenstein, Jr

R Markdown
The Definitive Guide
Yihui Xie, J.J. Allaire, Garrett Grolemund

Practical R for Mass Communication and Journalism
Sharon Machlis

For more information about this series, please visit: https://www.crcpress.com/go/the-r-series

Practical R for Mass Communication and Journalism

Sharon Machlis

CRC Press is an imprint of the
Taylor & Francis Group, an **informa** business
A CHAPMAN & HALL BOOK

CRC Press
Taylor & Francis Group
6000 Broken Sound Parkway NW, Suite 300
Boca Raton, FL 33487-2742

Printed on acid-free paper
Version Date: 20181204

International Standard Book Number-13: 978-1-138-72691-8 (Paperback)
International Standard Book Number-13: 978-1-138-38635-8 (Hardback)

Visit the Taylor & Francis Web site at
http://www.taylorandfrancis.com

and the CRC Press Web site at
http://www.crcpress.com

Dedication

To my parents, Barbara and Oscar Machlis, who gave me a great start in life, including instilling a love of learning. I miss you every day.

To my husband, Lee Gartenberg, for your love and unwavering support during countless hours of research, writing, and editing. Life is better because we're together.

And to the R community: Thanks to all of you who have written packages, shared other code, answered questions, and gone out of your way to create a welcoming and generous place. I hope to pay it forward.

Contents

Companion Web site

This book has a companion Web site at https://github.com/smach/R4JournalismBook (short link http://bit.ly/R4MassComm). The site includes data files used in most examples; links to more resources; and searchable tables of R tasks, functions, and packages presented in the book. You'll also be able to find corrections and updates there as needed.

Links to additional resources at the end of most chapters are also available online at https://smach.github.io/R4JournalismBook/booklinks.html. This is to save you from having to type out what are sometimes lengthy URLs if you have the paper version of the book.

Chapter 5 includes instructions on how to download the entire repository to your local computer from within R.

Chapter 1

Introduction

Imagine you've received a large spreadsheet with messy but important data, and you know it's got a story to tell. You spend lots of time cutting and pasting, writing formulas, and data "cleaning." One problem you're fixing is multiple versions of the same company's name: XYZ, XYZ Inc., X YZ Company, and so on. But when you're finally able to do your analysis, something looks wrong. And when you ask your source about it, he responds: Yes, sorry, there was a mistake in the data file. We'll send you a corrected spreadsheet shortly.

That was me a few years ago. After swearing (mostly to myself) in the newsroom, I had to re-create everything I did in that first Excel file in the second. The copying. The pasting. The formula-writing. The painfully long waits for formulas to execute. Spot checks of results. Standardizing on one version of each company's name so counts were accurate.

I vowed that wasn't going to happen again.

And I started learning R.

1.1 Why programming?

Being able to **easily repeat your own work** is an excellent reason to learn a programming language like R or Python when trying to make sense out of data. If you create a script containing all the steps for analyzing your data, you can easily re-do all your work. Simply execute a single command to run your code, and the script does the rest. That's true whether you've got a new, corrected data set or data that you receive regularly in the same basic format.

Do you receive monthly unemployment numbers that you process the same way? Daily arrest stats? Election results every year or two? Annual school test scores? Whatever it is you're analyzing, if you script it once, odds are you'll be well on your way to automating future analysis when the data changes.

But there are other advantages to using a script. This kind of workflow lets others check your work much more easily – what's known as **reproducible research** in the research community – than if you give someone a spreadsheet with multiple formulas. Even if a formula is correct in one cell, how can they – or you – be sure it's been properly copied and pasted? A script will run the same way each time it's executed.

There are certainly plenty of errors you can make when writing code. But one thing you *won't* have to worry about is whether you've copied and pasted or clicked and dragged properly (or used control-click instead of just click for certain types of advanced Excel formulas). And, someone else who's reviewing your work won't have to wonder either.

The good news: If you've written formulas in Excel, you've already done "programming" – just not on the command line, and not in a way that's easy to repeat.

The command line can seem intimidating at first for those who are used to working in more of a graphical environment. But after some practice, chances are you'll enjoy the power, flexibility, and what's-my-code-doing transparency of programming. And if you download the recommended RStudio software for writing and running your R code, you'll even have some benefits of a graphical user interface while creating your code.

1.2 Why R?

So, why R? One big attraction, especially for penny-pinching journalists and students, is that it's free and open source, unlike some powerful but pricey commercial platforms.

There are several popular open-source platforms for wrangling and analyzing data, and each has its ardent cheerleaders. If you've heard passionate arguments between iPhone and Android users, or Mac vs. Windows enthusiasts, you'll have a pretty good idea of what, say, R vs. Python arguments can sound like.

I don't want to disrespect Python, though – it's another great language. I happen to prefer R for much of my data work because *it was designed to analyze data.* And that means many of the things you want to do with data – structuring, summarizing, visualizing – are well thought out. There's a built-in data structure called a data frame that's spreadsheet-like in its organization, making it easy to apply calculations across columns or rows. And unlike most computer languages, R starts counting at 1 instead of 0, which means if you want row 273, you ask for 273 and not 272. (If you've never programmed before, you won't realize how unusual this is. If you have experience with one or more other languages, though, you may have to break yourself of some habits.)

It's fairly easy to install basic R and get started, whether on Windows or a Mac, which is something that can't necessarily be said about all programming languages.

R's capabilities are rapidly evolving, making it particularly interesting as a platform. The R ecosystem of today is far more robust than when I started learning R in 2012. For example, you can now create interactive Web maps and tables with just a couple of lines of code. It seems that every month, there are new, more elegant ways to wrangle, analyze, and visualize data.

Visualization is one of the most compelling features of R. When I did data exploration in Excel, I tended not to generate graphics until pretty late in my data work – usually only when I was ready to think about what chart to publish with a story. With R, though, it's easy to build dataviz into a standard workflow.

Finally, the large and growing community of R users is one of its best features. There are thousands of R "packages" – code written to enhance the core language or solve a specific problem – available for free download, making it likely that someone has already thought through how to solve a problem you might have with your data. And people in the community are usually eager to help if you run into problems.

1.3 Is this book for you?

There are three main audiences for Practical R for Mass Communication and Journalists:

- **Spreadsheet users who want to "graduate" to learning their first programming language.** If that's you, this book will get you gently up to speed so you become comfortable writing code to answer questions about your data.
- **People who know another programming language and now want to learn R.** While there will be some basic programming fundamentals discussed, this book focuses much less on theory and more on how to do useful work with R. So even though this is an entry-level R book, there should be plenty here to help you use R when dealing with data.
- **Communications professionals who already know some R but want to get some new tips and ideas for using R *in a newsroom or similar setting.*** If *this* is you, you may want to quickly

skim the next chapter on setting up R and RStudio. However, the rest of the book should help you learn ways to apply R specifically to the kind of work you do.

This book emphasizes "Practical" and "journalism/mass communications". There are already many good, generic R introductory books that go through language fundamentals. But I know that if you're a journalist, PR professional, political staffer/advocate or otherwise want to communicate ideas from data, you may not want to read a computer-science text as your first introduction to R. So, I don't start off with some basic information you'd typically get in a beginning R book, such as outlining different data types. Instead, I focus on the most important information you need to do useful work with R.

I want you to learn R with *data that you can imagine using in your newsroom, government office, or community group.* This book aims to show you how to use R in the real world – *your* real world. After the very basic introduction in chapter 2, theory and structure will come up mostly when needed, in situations you might actually encounter in your work.

We'll work together step by step to see how R can help you tell stories about topics like major weather events, election results, airline flight delays, and restaurant safety inspections. I took a lot of care when choosing projects and sample data in this book. I've seen enough Excel classes where journalists' eyes glaze over as an instructor drones on about which salespeople are eligible for bonuses. Compelling subjects are important in my work, and in yours.

Once you see how much R can help you when working with data, you may want to continue on your R-learning journey, perhaps with another book that focuses more on fundamentals. First, though, it's time to whet your appetite on what R can do for you.

Chapter 2

Get Started With R in a Few Easy Steps

2.1 What we'll cover

- Downloading and installing the software you need
- A tour of RStudio, including some useful tips
- Writing your first code
- Installing packages that add functionality to R
- Getting help

If you already have R on your computer and are comfortable using RStudio, you may want to just skim this chapter or even skip ahead to Chapter 3. Otherwise, follow along to get your system set up for R.

2.2 Download R and RStudio

You can download the most recent version of R at https://www.r-project.org/, which is the home of R (formally known as the R Project for Statistical Computing). The R-project home page usually includes information about the latest versions of R. Don't be put off by the sometimes odd nicknames for R versions, such as "Very, Very Secure Dishes" and "Bug in Your Hair" – the software is much more useful than you might assume from the nicknames. (The whimsical version names come from various Peanuts cartoons.)

There should also be a prominent link to download R. Click that download option and you should be taken to CRAN, the Comprehensive R Archive Network, and a list of CRAN servers, called mirrors, around the world. Pick a server and choose the *precompiled binary distribution* for your operating system. Once the file finishes downloading, install it like any other software program - run the .exe for Windows or .pkg for Mac.

You should be fine accepting all the Mac defaults. On Windows, you'll need to decide whether you want the 32- or 64-bit R version. (Unless you've got a pretty old system, chances are you'll want the 64-bit.)

This is all you need to start running R, but I strongly recommend also installing RStudio, a free platform designed to make it easier and more enjoyable to create and run R code. Head to RStudio.com and under products, look for RStudio and then RStudio *Desktop* (not Server), and download the free Open Source Edition version for your operating system. This, too, installs like a typical software program.

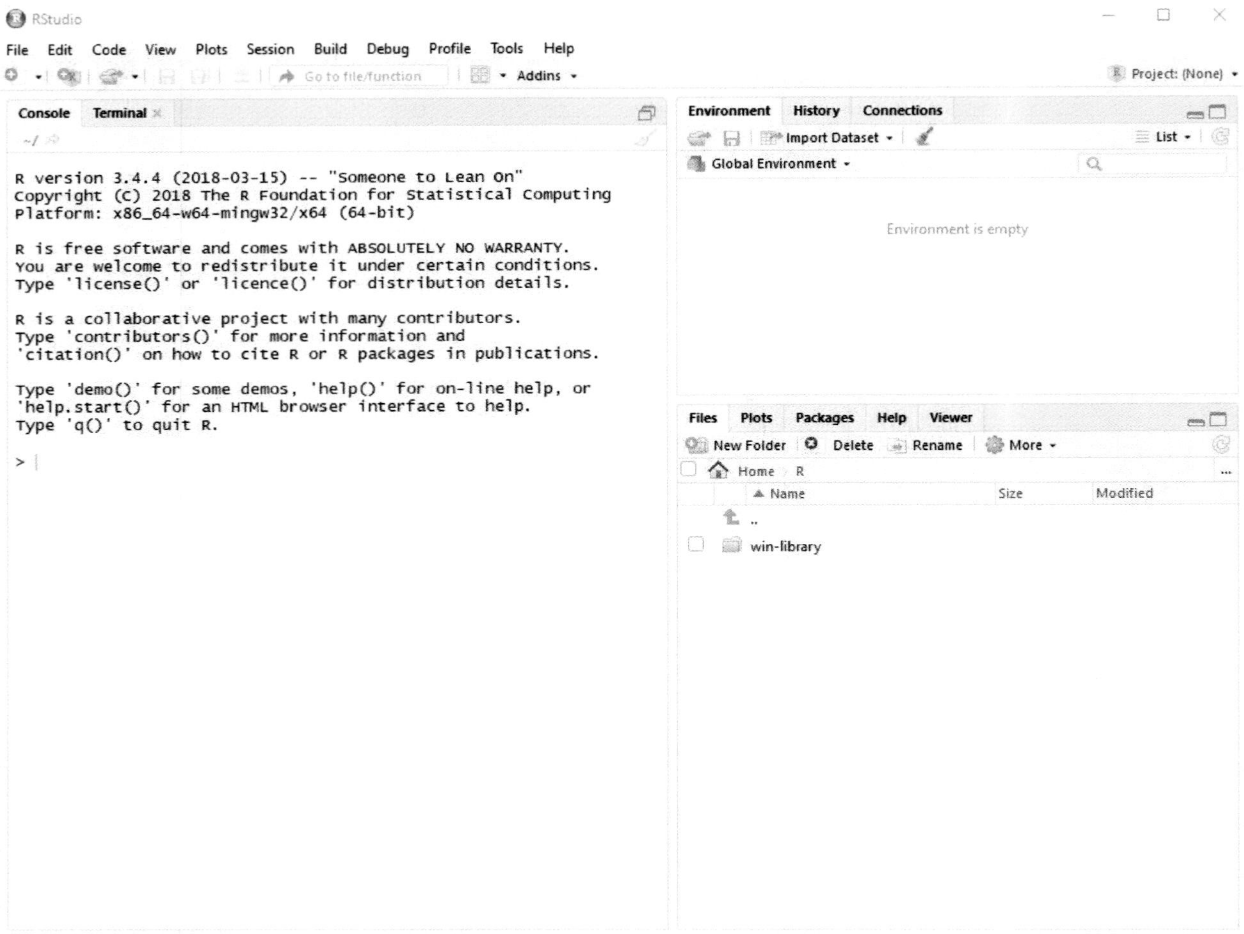

Figure 2.1: RStudio desktop software

2.3 A brief introduction to RStudio

2.3.1 The console

When you first start up RStudio, it will likely look something like Figure 2.1.

The area on the left is an interactive console, where you can type in commands and see the responses in real time. You can type in an arithmetic calculation such as `7 + 52` followed by the Enter key, or get the system date and time with `Sys.time()` and Enter.

Are you wondering why there's a period and parentheses in `Sys.time()`? Sys.time is an R *function*. The function happens to have a period in the function name; that dot doesn't have any additional significance the way periods do in some other languages. The parentheses after the function are more important - they mean you want to run Sys.time *as a function*. (If you type a function's name *without* the parentheses, R will show you the code behind the function instead of actually *executing* the function.)

If you click the up arrow on your keyboard while your cursor is in RStudio's console, the console will show the most recent command you've typed – quite handy if you want to repeat a command you just executed or modify an earlier command. Click the up arrow more than once and the console will show earlier commands. In addition, if you begin typing something in the console and hit control and the up arrow on Windows or command and the up arrow on Mac, you'll get a list of past commands you've typed that *start* with those characters. Control/command up arrow in the console at a blank line gives you a drop-down list of 20 or so most-recent commands.

Typing in a line like `7 + 52` is fine if you just want a quick calculation or two, but most of the time you'll want to be doing something more complex. If you go to File > New File > R Script in the RStudio menu, you should see another pane open on the left above the interactive console. This is where you can write a lengthier script with lots of lines of code, and save the file for future use.

Like in many Windows and Mac software programs, you can open multiple files in RStudio and each will have its own tab (which, just as in Excel or most browsers, can be dragged and dropped to re-order).

2.3.2 Other RStudio panes

The two panes on the right become useful as you create and run your code. In the top right pane, one tab shows your R Environment – what objects are loaded into your session at the moment. If you're new to programming, don't worry; this will make more sense once you start coding.

Another tab shows your command history. So if you typed `7 + 52` into the console, if you go to the History tab, you should see that in the history tab. You can select one or more lines in the history tab and then click the "To Console"" button at the top of the pane to send the line(s) back to the console, or the "To Source" option to send them to the top-left script pane.

You can search the history pane as well (you should see a search box at the upper right).

The lower right pane has several different, useful tabs. The first is a Files tab, similar to Windows File Explorer or Mac Finder. Although not quite as robust as those, this area is convenient for quickly renaming or deleting files, opening files, or changing your working directory.

The Plots tab is where you can view graphs and other data visualizations you create in R.

The Packages tab shows what packages are 1) available for you to use and 2) actually loaded into your working session. Anything listed is on your system; anything with a check mark to the left of the name is currently loaded in memory.

Finally, Help is where you can view help files for functions and packages. You'll likely be using that a lot, no matter how expert you become in R. If you click the home button on the help tab (it's the house icon), you'll see links to a lot of general R and RStudio information – some of it for beginners, some considerably more advanced. But you can also ask for more specific help in the R console for functions and packages, and search for help by keyword. (I'll show you how later in this chapter.)

2.3.3 RStudio Projects

RStudio is what's known as an IDE, or Integrated Development Environment. That's tech-speak for "software designed to make life easier for programmers." One common feature of IDEs is projects. In RStudio, opening a project automatically sets you up in the project's working directory, making it easier to find files stored in that directory.

Projects also keep track of which files you left open, so they'll still be open the next time you launch the software and open your project. Command history is different for each project as well, so you can scan through past commands that you typed just specifically for that project. There are some other useful features in projects as you get more familiar with the software.

You can create a new project by going to File > New Project. You'll be given three choices: New Directory (for a brand new project with nothing in it), Existing Project (to create a project from an existing directory that might already have files in it), and Version Control. If you've programmed before and are familiar with version control, you can create local files from a Git or Subversion repository. Project options at Tools > Project Options also let you easily create a version control repo for your project. I won't be covering version

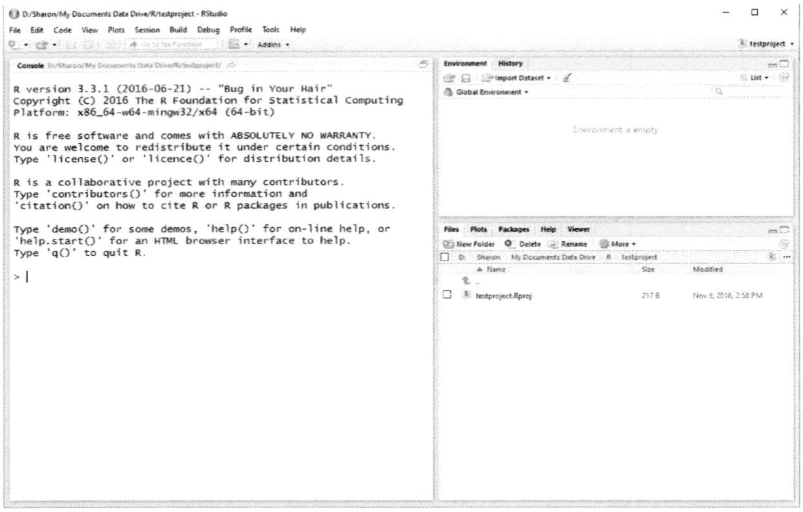

Figure 2.2: New project in RStudio

control in this book, but if it's of interest, there's some handy version-control integration within RStudio. (Jenny Bryan, formerly a professor at the University of British Columbia now with RStudio, has a nice roundup of using git with R and RStudio at happygitwithr.com.)

For now, create a new project with File > New Project. Select New Directory and then the first option, `Empty Project: Create a new project in an empty directory`. You'll be asked to name your new directory - perhaps call it something like testproject or R4CommBook. Leave "Create a git repository" and "Use packrat with this project" unchecked.

You should end up with an RStudio session that looks something like this:

Now it's time to write a little code.

2.4 Try out the console

2.4.1 Create your first object

Remember when we typed in 7 + 52? Perhaps next we'd like to first see the total of those two numbers and then calculate the average. We could first type

```
7 + 52
```

and then

```
(7 + 52)/2
```

But if we want to do more than one thing with the same data, it's best to store that data in a *variable.*

A variable is basically a container that stores some sort of value or values. In Excel, if you used a formula such as `=A1 + B1`, you used the variable A1 to mean "the value that's in cell A1" and B1 as "the value that's in cell B1." If the value in cell A1 changes, so will the value of `=A1 + B1`. In R (or any programming language), you can set a variable for a lot of different types of values. I can store the value 7 in a variable called num1 and store the value 52 in a variable called num2 like this:

```
num1 <- 7
num2 <- 52
```

That <- is R's "assignment operator." It just means num1 is assigned the numerical value 7, and num2 is assigned the numerical value 52. Most programming languages would use the equals sign, like `num1 = 7`. In fact, that will also work in R, but, well, it's not what the R cool kids do. Most R code you'll come across will use <-. Google's R style guide says use <-. So does Hadley Wickham, arguably the most well-known creator of packages in the R community. There are a couple of technical reasons for preferring <- to = , but for now, I hope you'll trust me that it's worth using.

Meanwhile, back to our variables.

Open up a new R script file in your RStudio test project by going to File > New File > R Script. Type

```
num1 <- 7
num2 <- 52
```

in the RStudio top pane with your new script file, and then save it. You can call it something like testscript.R – the .R file extension tells R and RStudio that it's an R script. You can save a file within RStudio by clicking on the disk icon, going to File > Save, or using the keyboard shortcut ctrl-S on Windows or cmd-S on Mac). If you forget to add .R to your file name, RStudio will add it for you, since you indicated when creating the file that it was an R script.

Now, you need to run your code. There are a couple of ways to run a script file within RStudio. To run the *entire file*, click on the Source button, not the Run button. The Run button only runs one line of code, whatever line your cursor is on. Ctrl-Enter on Windows and cmd-Enter on Mac will also run the single line of code your cursor is on (if a single command spans multiple lines, the full multi-line command will run).

To run several lines of code from within the RStudio script pane, highlight them (with the usual click-and-drag) and hit control-enter (command-enter on Mac).

If you'd like to re-run your entire script file every time you save it, click the "Source on Save" check box.

When you run the code, you should see num1 and num2 under "Values" in the Environment tab of the top right pane. That shows the variables exist and are available for use. If you type `num1` into the interactive console (bottom left pane) and hit enter, you'll get back the value of that variable, which in this case is 7.

If you type num slowly into the console and stop, you should notice that RStudio gives you several options: num1, num2, numeric (a base R function, meaning it's part of the core language) and a few more. This *autocomplete* can be quite helpful if you have a long variable name and want to make sure not to mis-type it. You can use your up and down arrows to select among RStudio's visible auto-complete suggestions and then hit tab to accept one. Enter works as well; hit Enter again if you're in the console and want to run it.

Now, instead of typing `7 + 52` to get the total, you can type

```
num1 + num2
```

```
## [1] 59
```

in the script file and run that line with ctrl/cmd-Enter.

The first line is what I typed into the console; the second line is my result. Any line beginning with `##` means that's what R returned. so it's not something you should type. The [1] on that results line means that is the first item of my results – not really needed when I've got one item, but helpful if there are 30 items over 5 or 10 lines.

I know that adding two numbers together isn't very exciting. But variables get much more useful once you have more data.

You'll often want to use variables to store results of your commands. Here, you can save the result of num1 + num2 in a variable called total1.

```
total1 <- num1 + num2
```

To see the *value* of the `total1` variable, either look at the environment tab in RStudio's top right pane, or type the variable name `total1` into the console.

A few important points about R variables:

- R is case sensitive, so total1 is a different variable from Total1 and TOTAL1.
- Variable names can start with a letter or a period (if a period, the next character can't be a number) and can then contain letters, numbers, periods, and underscores. `my_total` is a valid variable name but `_mytotal` is not.
- Variables can hold different types of data, not just a single number. A variable can hold many numbers at once, a character string, and other things (this book will cover several of the most useful data types, but not all of them). Some programming languages are pretty strict about data types, but R isn't; it's quite easy to change the type of data a variable holds, and it's not necessary to specify a data type in advance.
- There are some reserved words that can't be used as variable names, and you probably don't want to use existing function names as variables even when allowed, since that can screw up your access to a function you may need later. I often use names like `mytotal` or `myaverage` as function names in my own work to make sure I'm not stepping on an existing function name.
- An R variable is technically an *object*. That means it has certain characteristics depending on what kind of data it holds.

2.4.2 Data types you're likely to use often

Storing a single number in a variable can occasionally be useful – think of doing currency conversions and storing how many dollars you can buy with a euro – but a lot of times, you'll want to work with data that's a bit more robust. Here are some of R's options.

Character strings. You assign a character string to a variable by putting the data in quotation marks. Typically double quotes are used, such as `mystring1 <- "Hello, world!"`, although single quotes work as well.

Vectors. It's probably easiest to show a vector than to explain it. 2,4,6,8 is a vector of integers; "a", "b", "c", "def" is a vector of character strings. A vector can only have one type of data - all integers, all strings, all logical TRUE or FALSE, and so on. If you try to mix types in one vector, R will turn all your data into a single type – 1, 2 and "three" will end up as character strings "1", "2" and "three". (If you want to mix data types, you need to use another type of R object called a list. Lists are useful but can get complicated. We'll cover them later.)

You create a vector with R's c() function, which you may want to remember as being either short for "concatenate" or "combine". `mynumbers <- c(2,4,6,8)` creates a vector of integers; mystrings <- c("a", "b", "c", "def") creates a vector of character strings.

Remember before, when I said R objects have certain characteristics? Vectors have characteristics, or attributes, such as `length` and whether each item has a `name`; you can also test whether or not it's a vector:

```
mynumbers <- c(2,4,6,8)
length(mynumbers)
```

```
## [1] 4
```

```
is.vector(mynumbers)
```

```
## [1] TRUE
```

```
names(mynumbers)
```

```
## NULL
```

Reminder: The two pound signs in front of the results of these commands indicate that they're responses from the console and not something you should type in yourself. You won't see the `##` when working in your own console. You *will* see the [1] before results of length() and is.vector().

length(), by the way, counts the number of items in a vector or R list but *not* the number of characters in a string. It might seem logical that `length("cat")` would return 3 – the number of characters in the string "cat" – but it won't. `length("cat")` is *1*, because "cat" is 1 item. `length(c("cat", "dog"))` would be 2, because `c("cat", "dog")` is a vector with 2 items.

If you want to count the number of characters in a string, use `nchar()`.

You can do arithmetic on a vector of numbers. `mynumbers / 2` will divide *each item* in the vector by 2:

```
mynumbers / 2
```

```
## [1] 1 2 3 4
```

Data frames. This built-in data type is one of the most compelling things about working in R. A data frame is somewhat like a spreadsheet: It is 2-dimensional and has rows and columns. If you read a spreadsheet or CSV file into R, it will usually come in as a data frame.

We'll start analyzing and visualizing data in data frames very soon.

2.5 Install packages

As I mentioned earlier, one of the great benefits of R is its community of users and the many packages they're willing to share with the rest of us. It's time to start installing a few.

There are a couple of main sources for R packages. One is CRAN, the Comprehensive R Archive Network. Packages must go through an approval process to be listed on CRAN, and can be installed right from the RStudio command console with `install.packages("NameOfPackage")`. A lot of other R packages are available on GitHub, a developers' platform for sharing code. We'll be installing from both sources.

First, some of my favorite CRAN packages. There's a group of packages designed to work together that Hadley Wickham named the "tidyverse," called that because they aim to create and analyze so-called "tidy" (well-structured) data. A number of people both within and outside of RStudio have since contributed to these packages. You probably won't need everything in the tidyverse to run code in this book. But with a single install command, you'll get several of my favorites, including dplyr for data wrangling and ggplot2 for visualization. Install the tidyverse with

```
install.packages("tidyverse")
```

You may be asked to pick a CRAN mirror (if so, select one from the list) and a location for your package library on your system. This installs more than two dozen R packages, so it may take awhile. But you should only need to do this once for each system you use to run R.

Another package I suggest you install now is rio:

```
install.packages("rio")
```

for importing various types of data files into R.

And finally, please install the pacman package, which I'll be using in most of the book to make sure you've got the proper packages installed and loaded for each chapter.

```
install.packages("pacman")
```

We'll be installing other packages later on.

2.6 Additional infrastructure

When building some robust R packages, you may need additional software infrastructure. To save yourself the hassle of running into this problem later on, I'd also suggest the following installations (although feel free to skip those for now if you feel like you've done enough system administration for one sitting):

On Windows: Install Rtools. Download the appropriate Rtools.exe executable file for your version of R here: https://cran.r-project.org/bin/windows/Rtools/

On OS X: Install Xcode, Apple's development environment, which should be available in the App Store.

2.7 Getting help with packages and functions

How do you learn to use a package or know what functions it contains? A well-written R package includes a list of all its available functions as well as help files for every function. For a package's main help file, type `help(package="NameOfPackage")`. You can try it now with dplyr: `help(package="dplyr")`. You should see the main dplyr help file appear in RStudio's lower-right pane. There's a list of available functions, and you can click through each one to get more details about the function.

Some packages also have what are called vignettes, which are short articles explaining how to use the package or do specific tasks. If you don't see a link to "User guides, package vignettes and other documentation" on the main help page, you can try `browseVignettes("NameOfPackage")`. The command `browseVignettes()` without anything inside the parentheses will show you all available vignettes on your system. Note: The more technical term for what's between the parentheses when you run a function is *argument*.

For dplyr, run `help(package = "dplyr")` and click the link to User guides, package vignettes, and other documentation in the main help page. Select the dplyr::dplyr vignette (dplyr::introduction in some earlier versions of the package). There, you'll see a nice overview of the package as well as how to use several of its most important functions.

To see a function's help file without first loading the package's main help file, you can type `help("NameOfFunction")` or just `?NameOfFunction` (remember that function names are case sensitive). The package needs to be loaded into your session with `library("NameOfPackage")` for the ? shortcut to work. If the package isn't loaded, `help(NameOfFunction, package = "NameOfPackage")` will work.

In addition, if you've typed the name of a function (either in the console or script pane) from a package that's loaded into your working session (or from base R), just hit the F1 key. Your cursor has to be *just after the last letter of the function name for this to work*, so if you've already typed, say, arrange(), move your cursor to just after the e in arrange and F1 help should still work; if your cursor is in the middle of the parentheses, it won't. I was using RStudio for several years before I discovered this time-saver and stopped jumping from my script window to the console to type `?NameOfFunction` to see a function's help file.

Finally, if you want to search for a function but either aren't sure which package it's in or don't want to type out the lengthy `help(NameOfFunction, package = "NameOfPackage")` when its library isn't loaded, you can use the double question mark. `??median` will bring up links to all functions that contain median in either their name or description for all packages on your system, whether or not they're currently loaded into memory. This will also do a little partial matching, bringing up results that almost match your search term.

When you're ready to close RStudio, you'll be asked if you want to save any unsaved files – you probably do, even though RStudio will keep your project's unsaved files as is – and whether you want to save your workspace image to an .RData file. In general, I'd advise *not* to save your *workspace*, because all of your variables will be stored and re-loaded the next time you launch RStudio. It's too easy to forget about previously stored variables that can interfere with later work, not to mention taking up memory with things you might not need. Best to start off clean each session. If you're performing lengthy calculations that you don't want to repeat, there are ways to save variables to disk without auto-loading your entire workspace.

If, like I am, you're sure you don't want to be saving your workspace, you can tell RStudio to stop asking you. Go to Tools > Global Options and choose Never from the drop-down list for `Save workspace to .RData on exit`.

2.8 RStudio keyboard shortcuts

You can see a list of RStudio keyboard shortcuts by going to Tools > Keyboard Shortcuts Help or using the keyboard shortcut alt-shift-k. You can customize keyboard shortcuts for RStudio menu items (and external addins) at Tools > Modify Keyboard Shortcuts.

2.9 Additional files available online

This book has a GitHub repository at https://github.com/smach/R4JournalismBook. I'll give you detailed instructions for downloading all the files in Chapter 5. But if you'd like to take a sneak peek before then, you can take a look now. Don't worry if you're not familiar with GitHub, or the git system it's based upon. A "GitHub repository" is just a software project that includes a robust way of tracking changes in files, making it easy for many people to work on a project at the same time. You won't need to know specifics of how git and GitHub work in order to download and use the book's files.

2.10 Wrap-Up

We went over a lot of ground here to cover R basics! We downloaded and installed both R and the RStudio IDE. We learned our way around RStudio's panels, including the interactive console at the bottom left, script console at the top left, environment tab at the top right, and several tabs at the bottom right including Files, Plots, Packages, and Help.

We stored data in variables and learned some different R data types. Basic "atomic" data types include numbers and character strings. What we can think of as data structure types include vectors and the extremely handy, sort-of-spreadsheet-like data frames.

We learned how to install packages from the official CRAN repository with `install.packages("PackageName")` and how to get help with packages and functions using `help()`, `?`, and `??`.

We installed the `tidyverse`, `rio`, and `pacman` packages using install.packages().

And finally, we went over some useful RStudio tips, such as control-upArrow to see a list of past commands.

Next up: With some R basics under your belt, it's time to have a little fun.

2.11 Additional resources

- **My Computerworld Beginner's Guide to R.** Available on the Computerworld website at http://cwrld.us/IntroToR or as a PDF download at http://cwrld.us/RGuidePDF (free registration required for the PDF).
- **RStudio Tips on Twitter** Follow the account @rstudiotips. You can scan prior tips at https://twitter.com/rstudiotips.
- **RStudio IDE Easy Tricks you Might've Missed.** Some useful tips from RStudio at http://bit.ly/RStudioMoreTips.

- ** Installation help.** No matter how many step-by-step instructions a book includes, there may be times when software won't install properly on someone's computer. It's simply not possible to anticipate every problem, whether conflicts with other installed programs or firewall/network difficulties.

If you're having issues, you may want to try searching for similar error messages. You can start off with Google, but if that doesn't work, try searching on the developers' site Stack Overflow at stackoverflow.com – include the R language tag [r] in your search – or RStudio's community https://community.rstudio.com.

If you need to post a question, give as many details as you can. For example, "I'm trying to install package X and it's not working" doesn't give readers any idea what's going wrong. "I'm getting the following error message when trying to install package X with the code install.packages("PackageX") on my Windows 10 system running R version 3.4.4:" along with the error message would be better.

Hopefully, though, you won't need this kind of help at the outset of your R journey!

Chapter 3

See How Much You Can Do in a Few Lines of Code

The sole purpose of this chapter is instant gratification. Fundamentals are important, but so is seeing a few of the cool things you can do in R *with very little code.* I do realize that adding 7 + 52 in a console, as I demonstrated in the last chapter, isn't all that compelling. I want to get you excited about R's potential! But if you're the type of person who gets frustrated doing things without fully understanding them, you may want to skip ahead to Chapter 4 and come back here later. This chapter is designed to show you some eye candy, not to give detailed explanations.

3.1 Packages needed in this chapter

(I'll include installation code when required.)

- quantmod
- dygraphs
- htmlwidgets

3.2 What we'll cover

- Quick interactive graphs with the quantmod and dygraphs packages
- Interactive maps with a few lines of code.

You'll first need to install two packages, which you can do by running this code in your interactive R console (bottom left pane):

```
install.packages(c("quantmod", "dygraphs"))
```

quantmod is a library for financial analysis. dygraphs creates interactive Web graphics of data over time.

A note about installing packages: Usually, this is pretty seamless in R. Occasionally, you might get a message that the package you're trying to install *won't* install, because another package is missing. R *should* automatically install needed packages by default, but sometimes there's a glitch. If you get an error message that some other package you didn't know about is missing, try installing *that* one manually with `install.packages("missingPackageName")` and then run the install on the package you want.

In addition, sometimes you may be asked whether you want to "install from sources the package which needs compilation?" Most of the time, that just means, "Would you like the absolute latest version that doesn't have a handy, single download file yet?" For all packages we're using in this book, choosing n for no is easier and should work just fine.

Once installation is done, load both packages into your current R session with

```
library("quantmod")
library("dygraphs")
```

Finally, if you haven't yet set up an RStudio project for your code related to this book, as I described in the last chapter, create one for this work by going to File > New Project.

Now, let's give some those packages a try.

3.3 Simple stock market graphing

How did Google stock do since the 2008 stock market crash? Let's take a look ... with these two lines of code:

```
google_stock_prices <- getSymbols("GOOG", src = "yahoo", from = "2008-01-01", auto.assign = FALSE)
chartSeries(google_stock_prices)
```

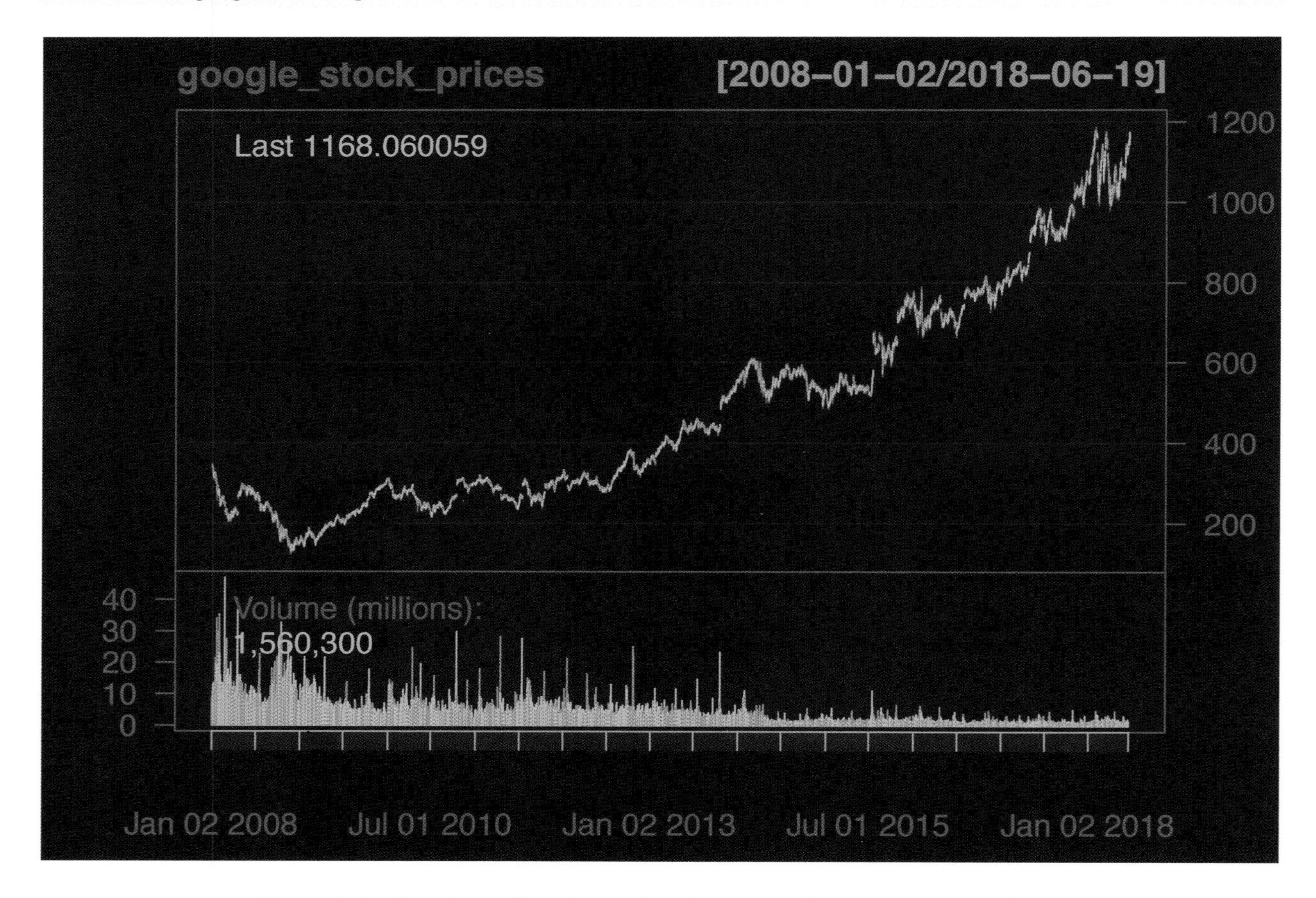

Figure 3.1: Graphing Google stock prices with the quantmod package.

Setup notes: You can type each line of code into the console, but it will be more convenient in the long run to set up a script file for this chapter. Go to File > New File > R Script (or use the keyboard shortcut Ctrl-Shift-N on Windows / Cmnd-Shift-N on Mac). Type the commands into the script file, save it with the disk save icon, File > Save, or the ctrl/cmnd-S keyboard shortcut.

To run all the code in the file, you can click the Source link above and to the right of the script panel (which is the top left panel). To run a few lines of code, select them the usual way (click and drag with your mouse) and hit control-Enter (command-Enter on Mac). To run one line of code, put your cursor anywhere on that line and hit control-Enter or command-Enter. When you do that, your cursor will jump to the next line of code and you can hit control/command-Enter again to quickly run the next line of code.

getSymbols() is quantmod's powerful function for getting historical financial data from the Internet and importing it directly into R. `src = yahoo` tells quantmod that I want to pull the data from Yahoo; I could also have specified google to import from Google Finance. `from` tells quantmod when to start pulling the data. And auto.assign=FALSE deals with a quirk in older versions of quantmod, so it behaves like most R functions and can store functions in an R variable with the `<-` assignment operator.

The dygraphs package can make this graph interactive, so that you can mouse over the line and see underlying data, as well as click and drag to zoom in on a portion of the graph. It will look something like Figure 3.2. Again, just one line of code, this one specifying "I just want the GOOG.Close column (with closing prices) and a title of Google Stock Price Starting in 2008".

```
dygraph(google_stock_prices[,c("GOOG.Close")], main = "Google Stock Price Starting in 2008")
```

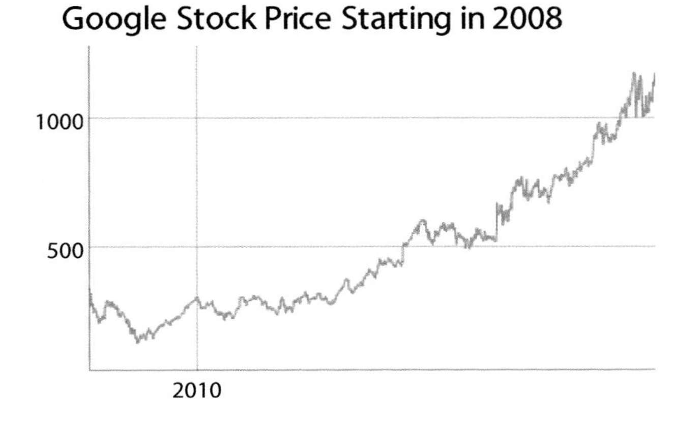

Figure 3.2: Graphing Google stock prices with the dygraphs package.

3.4 Download and graph a city's median income

quantmod includes a function that lets you import data directly from the U.S. Federal Reserve – more specifically the St. Louis Federal Reserve's FRED database.

I went to FRED at https://fred.stlouisfed.org/ and searched for "median household income for San Francisco." This wasn't to get the data, but to find out which table contains the data. The URL for FRED's "Estimate of Median Household Income for San Francisco County/City, CA." was https://fred.stlouisfed.org/series/MHICA06075A052NCEN. The character/number string after fred.stlouisfed.org/series/ was what I needed; it's the St. Louis Fed's symbol for this data.

You may want to do the same search, so you can copy the MHICA06075A052NCEN portion of it into your clipboard from the FRED url instead of typing it manually.

Now try running the following code (I'll explain it in a bit) for some instant R gratification (gRatification?). Being able to paste MHICA06075A052NCEN from your clipboard should make this less onerous.

```
sfincome <- getSymbols("MHICA06075A052NCEN", src="FRED", auto.assign = FALSE)
names(sfincome) <- "Income"
dygraph(sfincome, main = "San Francisco Median Household Income")
```

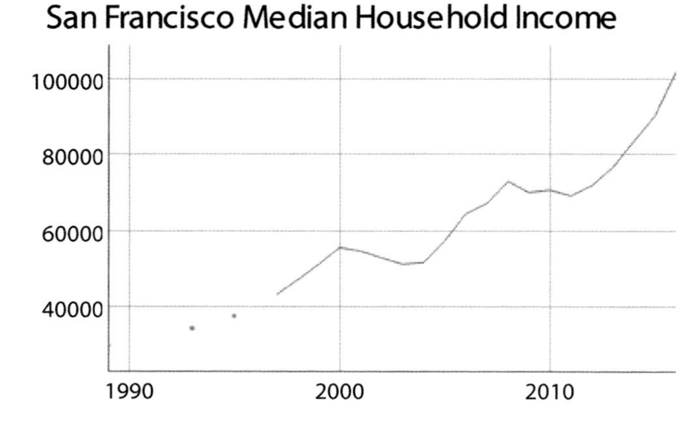

Figure 3.3: Graphing San Francisco household income with the dygraphs package.

You should see a graph that looks like Figure 3.3 in the Viewer tab of RStudio's lower right pane.

The zoom button above the graph on the left will open the pane into a larger window. The icon showing an arrow pointing to the upper right will open the graph in a browser window (in Figure 3.4, it's the icon directly over sco in San Francisco).

The graph is interactive. If you move your mouse from left to right anywhere on the graph, you'll see information about the closest underlying data point in the top right of the window. And if you click and drag from left to right inside the graph to define a portion of it - say from 2000 to 2010 - the graph will zoom in. (Double-clicking zooms out again, or you can always refresh the page.) This graph can be saved as an HTML "widget" or exported to a static JPG or PNG image using the RStudio Export menu item in this Viewer pane.

To save this graph as an HTML widget, first store it in an R variable, such as `sfmediangraph <- dygraph(sfincome, main = "San Francisco Median Household Income")`. Next, install the htmlwidget package with `install.packages("htmlwidgets")` and load it with `library("htmlwidgets")`. Finally, use the `saveWidget()` function to save your sfmediangraph object to a file, such as

```
saveWidget(sfmediangraph, file="sfmediangraph.html")
```

Find the sfmediangraph.html file on your computer and open it in a browser. You should see the same interactive HTML graph as you saw in RStudio, but it is now reusable on another Web site.

If by any chance you're thinking "This looks like a Web graphic I could make with JavaScript," you're right. The dygraphs package is an R wrapper for a JavaScript library, also called dygraphs. But with the R package, you can generate dygraphs JavaScript completely in R.

Now, the explanation of the code that I promised:

The first two lines load the quantmod and dygraphs packages into your current working session.

The third line uses quantmod's getSymbols() function to pull data from an external source. The function takes two "arguments" – options that the function needs to do its job. That first argument, "MHICA06075A052NCEN", is the symbol (the one we looked up in FRED) for retrieving data we want. The

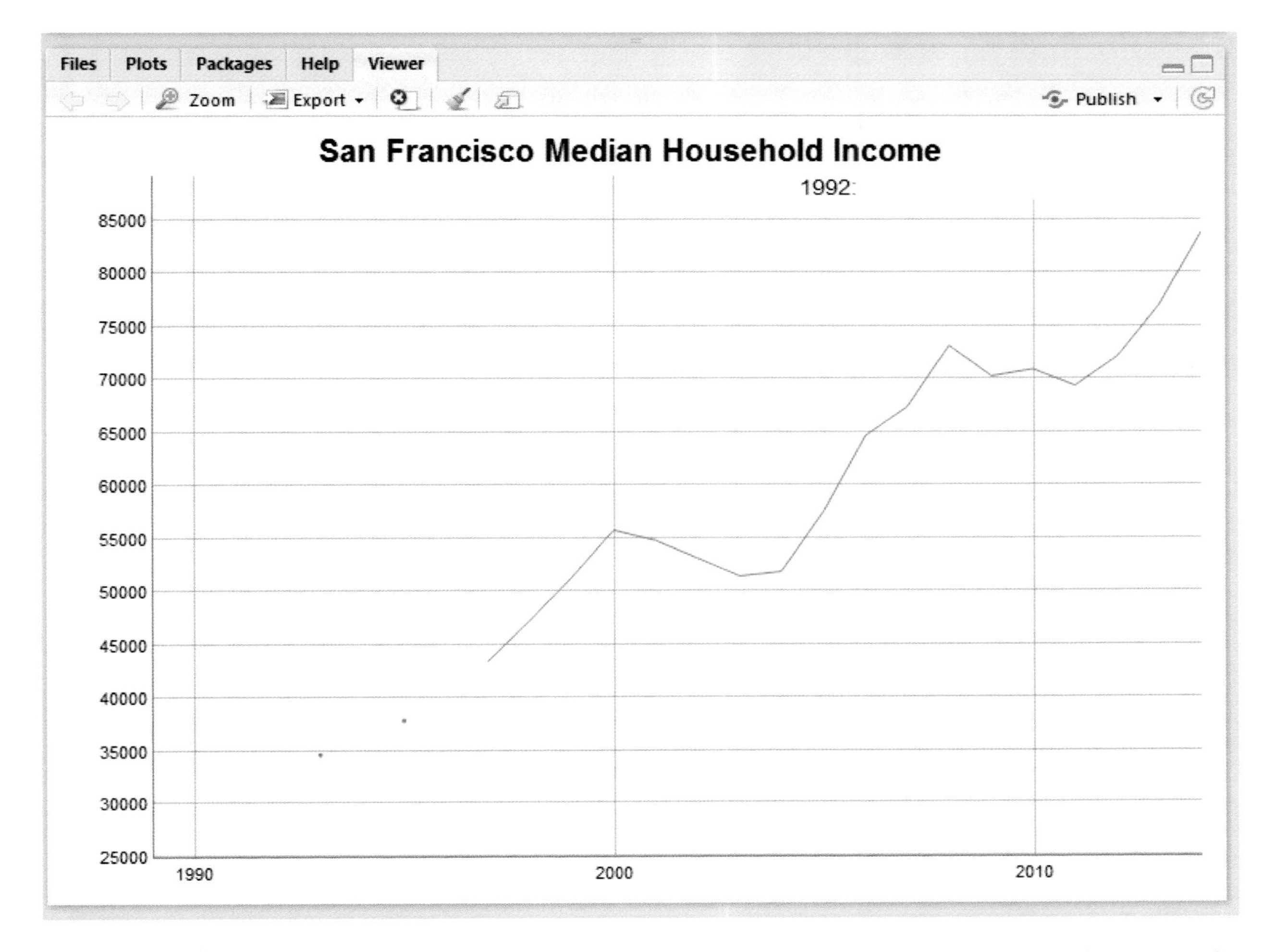

Figure 3.4: The zoom button at the top left will let you open the graph in a larger window; the icon to the right of the broom opens it in a browser.

second argument, src="FRED", sets the data source to FRED, the St. Louis Federal Reserve's database. And you saw "autoassign = FALSE" in the previous section.

If you check the structure of this sfincome object with str(sfincome), you'll see that it's a special type of R object dealing with time series. Data starts in 1989, and those NA data listings you see stand for "Not Available," meaning that some points are missing (which you probably noticed on the graph).

The name of the lone data column with the median-income data is "MHICA06075A052NCEN". The data column title shows up on the graph, and I don't like that character string as my data label. So, in the fourth line, I change the name of that data column to "Income" with `names(sfincome) <- "Income"`. You can also run names() on an R object *without* the assignment operator, such as `names(sfincome)`, to see existing names.

(If you're wondering why no name for the *date* column shows up, that's how R time series are structured.)

The fifth line creates the graph, using the dygraph() function from the dygraphs package. The first argument, sfdata, tells dygraph what data set to graph. The second argument, main, is just the headline for the graph.

3.5 So many packages!

At this point, you might want to ask me: How did you know about these packages? How did you know what functions to use, and how they work? And, as you learn about different packages, how do you remember which ones to use when?

All good questions (and ones I'm often asked at workshops). I try to keep up with package developments by looking at top tweets with the #rstats Twitter hashtag, scanning a number of R blogs, and following posts on the Google Plus R group. This may be easier for me than a lot of other R users because 1) I cover developments in data analysis for my day job as a tech journalist, and 2) I'm somewhat R-obsessed. If you don't follow R developments for *your* job, a good shortcut to keep up with the latest R developments is the community-sourced R Weekly, which tries to round up the most interesting R news in a fairly easy-to-scan post at rweekly.org. You can also follow my tweets with the #rstats hashtag from my @sharon000 account.

When I learn about a very useful package, I add it to an interactive, searchable table published by Computerworld, which is available at http://bit.ly/Rpackages. You might want to keep your own spreadsheet of favorite packages and functions, whether or not it's published for others to read. I'd suggest keeping it somewhere in the cloud even if it's not public, like in a Google sheet or Excel spreadsheet on OneDrive, so you can access it from different computers.

There's another way to store some of your favorite functions right in RStudio, called code snippets. I'll be covering them in Chapter 6.

After you discover a package, reading the introductory vignette can help you figure out how to use it. Also, even if a package is on CRAN, the code may be on GitHub as well, and package authors will often add useful information there. In fact, there's an extremely helpful tutorial on the dygraph package at https://rstudio.github.io/dygraphs/. If you Google "R" and a package name you may come across other useful content (when I wrote this, the RStudio dygraphs tutorial on GitHub came up first when Googling R dygraphs).

3.6 Running functions without loading packages

Returning to the project at hand, I'd like to show you a slightly different way to generate the same graph:

```
sfincome <- quantmod::getSymbols("MHICA06075A052NCEN", src="FRED", auto.assign = FALSE)
names(sfincome) <- "Income"
dygraphs::dygraph(sfincome, main = "San Francisco Median Household Income")
```

In this code above, I use `quantmod::getSymbols()` instead of just `getSymbols()` and `dygraphs::dygraph()` instead of `dygraph()`. By adding `packagename::` before a function, I'm able to access a function from an external package without having to load the package into memory first with library(packageName). This syntax `PackageName::FunctionName` works for any package that *exists on your computer's hard drive but isn't loaded into memory.* There can be a couple of advantages to this. First, if the package is large and you're only using one function once, you can save system memory by not loading all the package's functions into your session. (However, if you're using several of a package's functions multiple times throughout a script, it's often easier to just load it into memory.)

Second, when you look at a code snippet months later, it's clear which package each function belongs to.

Finally, if you've got multiple external packages loaded into memory, it's possible that the author of package1 named a function the same thing as the author of package2. Using package1::myfun() lets R know you want the myfun() function in package1, and not any other function named myfun() from some other package. We'll encounter that exact situation with two different functions called `describe()`.

3.7 Comparing one city's data to the US median

Whether you're a journalist or government staffer looking at a trend line like the rise in San Francisco median household income this, it's helpful to keep a key statistical question in mind: *compared to what?* A 6% increase in a city's median household income over several years might be impressive if overall national income stayed flat in the same period, but could be viewed as sluggish if the US household median rose 10% over that same period.

So let's add national median income to the graph. I looked up US income on FRED, and it's "MHIUS00000A052NCEN". Get the national data with

```
usincome <- getSymbols("MHIUS00000A052NCEN", src="FRED")
names(usincome) <- "US Income"
```

and then add the national data columns to the San Francisco data set with base R's `cbind()` function. cbind means "bind" two data sets together by adding one data column to another, side by side. You can also `rbind()` to add rows from one data set below another.

```
mygraphdata <- cbind(sfincome, usincome)

## Warning in merge.xts(..., all = all, fill = fill, suffixes = suffixes): NAs
## introduced by coercion
```

Now, use the `dygraph()` function to graph the mygraphdata data set the same way you created the first graph:

```
dygraph(mygraphdata, main = "Median Household Income")
```

One of the best things about scripting this: When new data is available from the Fed, you can just run the same code and you'll have an updated graph! Now, imagine how useful this is if you work with data that updates monthly or weekly.

Another advantage: Once you've got the basic code for pulling data from the Fed, it's easy to change to another data point such as unemployment. The code for US unemployment data on FRED is "UNRATE" and San Francisco is "SANF806UR", so you can just swap those in for "MHICA06075A052NCEN" and "MHIUS00000A052NCEN" and get your data set:

```
sfunemp <- getSymbols("SANF806UR", src="FRED", auto.assign = FALSE)
names(sfunemp) <- "SFRate"
usunemp <- getSymbols("UNRATE", src="FRED", auto.assign = FALSE)
names(usunemp) <- "USRate"
unemploymentdata <- cbind(sfunemp, usunemp)
dygraph(unemploymentdata, main = "Monthly Unemployment Rates, US and San Francisco")
```

Because San Francisco unemployment data is only available since 1990, it could be better to just show all data from 1990 onward instead of displaying earlier national data. Time series have their own unusual way of subsetting data; this code

```
unemploymentdata <- unemploymentdata["1990/"]
```

will update the unemploymentdata variable so it contains only information from 1990 and later.

Now you can draw the graph with:

```
dygraph(unemploymentdata, main = "Monthly Unemployment Rates, US and San Francisco")
```

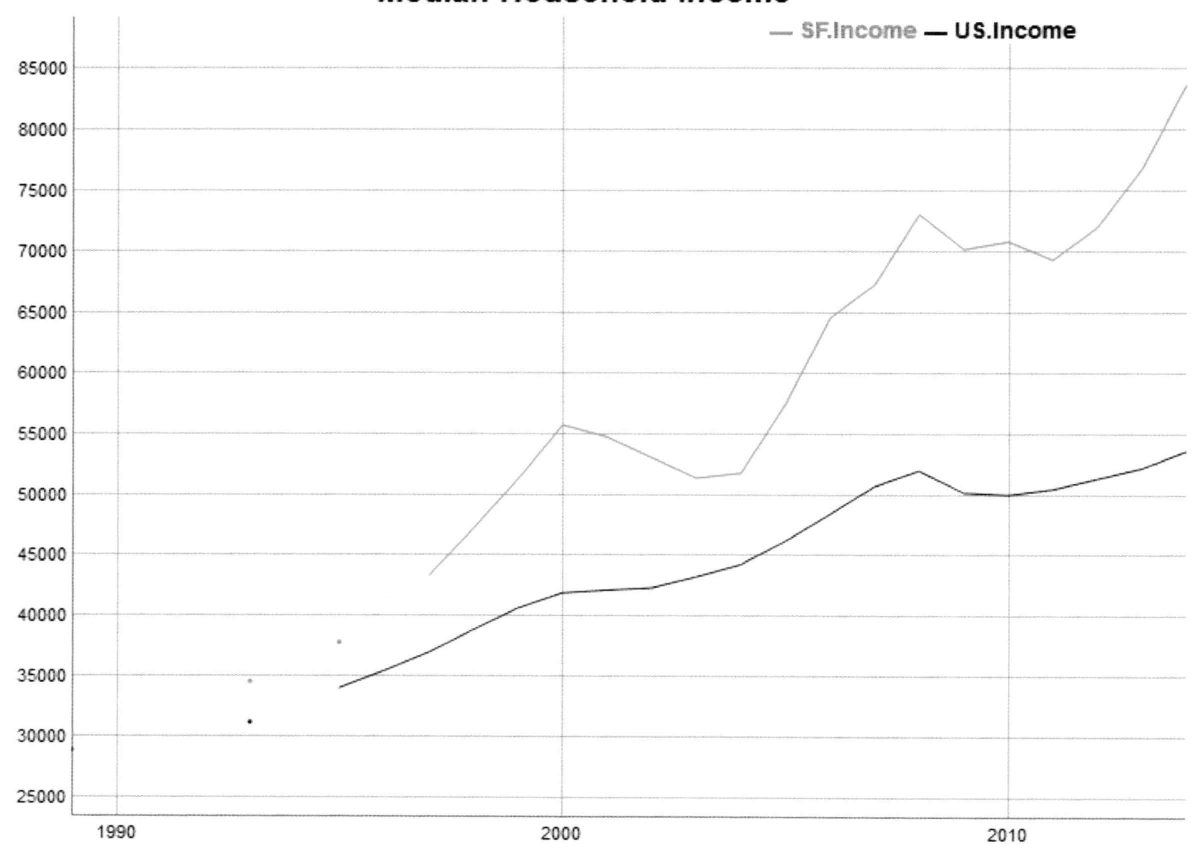

Figure 3.5: Graph of San Francisco and US median household income

3.8 Run a remote script to make an interactive map

I've posted most of the data files from this book in a public repository on GitHub. Soon we'll create a new RStudio project for all these files. For now, though, I'd like you to run one file from that "repo" at https://raw.githubusercontent.com/smach/R4JournalismBook/master/03_map.R . If that's somewhat onerous to type, go to the bit.ly link http://bit.ly/R4JournalismLinks and find and copy the URL under the *Chapter 3* heading.

Then run

```
source("https://raw.githubusercontent.com/smach/R4JournalismBook/master/03_map.R")
make_mymap()
```

and see what happens. Eventually, you should see an interactive map in your bottom-right RStudio Viewer pane, showing a few Starbucks in Atlanta, like the one in Figure 3.6. This is a national map, though – you can zoom out and see other areas of the U.S.

Before you use this to find a coffee shop near you, though, be warned that the latest available data file for Starbucks locations is from 2012. The point is to demonstrate making a cool interactive map with pop-ups, and I've found that the Starbucks data set tends to resonate in a crowd of journalists.

I also wanted to show you how to include code from an external file in your scripts. `source("myfile")` lets you run code from an external file in your current script, whether that file is somewhere else on your own

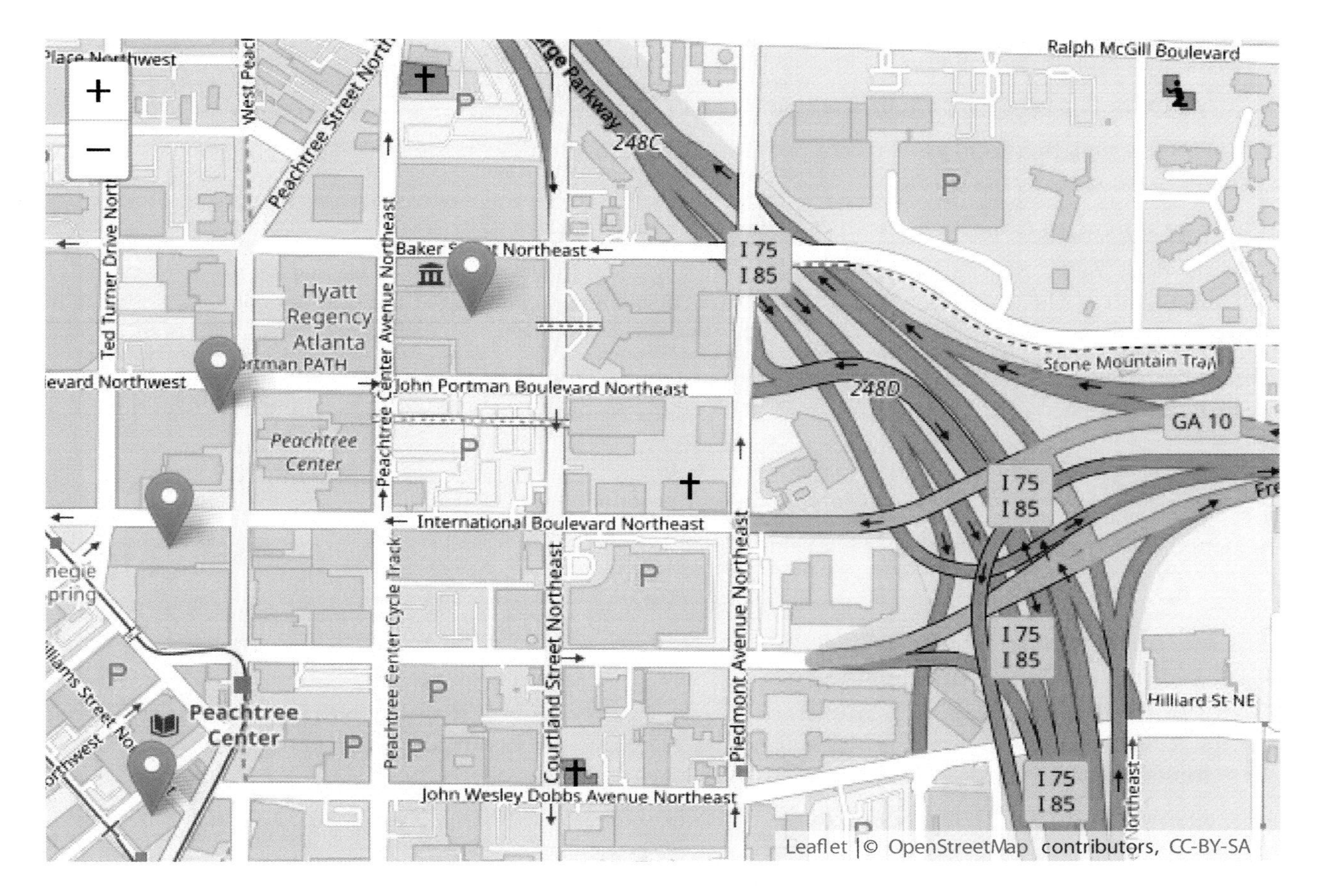

Figure 3.6: Interactive map of Starbucks locations.

system or a file that you've stored remotely.

3.9 Bonus map: Mapping income data

We'll be doing a lot more work with maps later. Meanwhile, though, if you'd like a sneak peek, try running the following remote code. Note that it will be loading several packages on your system the first time it executes as well as a data file, so it may take a little while to execute.

You can find and copy that lengthy URL by heading to http://bit.ly/R4JournalismLinks and copying the second URL under the Chapter 3 heading.

```
source("https://raw.githubusercontent.com/smach/R4JournalismBook/master/03_manhattan_income_map.R")
get_household_income_in_county()
```

Click on different portions of the map to get pop-ups with data details.

3.10 Wrap-Up

I hope that's gotten you enthused about some of the power of R. This chapter covered a lot of code somewhat quickly, but the point was to see that R can do some cool things, not to learn and understand every nuance of programming presented here.

Here's what's worth remembering from this chapter:

Load an entire R package into your working session's memory with `library(mypackage)`.

You can use a function from an external package *without* loading the entire package into memory, using the syntax `mypackage::myfunction()`.

Run R code from an external file with `source("myfile.R")`.

Combine data frames by column with `cbind()` (putting them together side by side) and by row with `rbind()` (putting them together one below the other). We'll cover more sophisticated merging by common columns in later chapters.

`names(myobject)` will display an object's existing names. You can also *change* the names of one or more items in an R object with names(myobject), such as names(mydataframe) <- c("Column1", "Column2").

NA stands for Not Available and indicates missing data in R.

There are some very elegant ways of importing and visualizing data with R.

Next up: How to import all sorts of data into R.

3.11 Additional resources

- **5 data visualizations in 5 minutes:** Each in 5 lines or less of R http://bit.ly/5LinesOrLess. This is a version of my 5-minute lightning talk at the 2015 National Institute for Computer Assisted Reporting conference. With video.
- **htmlwidgets for R** htmlwidgets.org. Find out more about interactive HTML for R including dozens of packages and a showcase of examples.

Chapter 4

Import Data into R

Before you can analyze and visualize data, you have to get that data into R. There are some different ways to do this, depending on how your data is formatted and where it's located.

Usually, the function you use to import data depends on the data's file format. In base R, for example, you can import a CSV file with `read.csv()`. Hadley Wickham created a package called readxl that, as you might expect, has a function to read in Excel files. There's another package, googlesheets, for pulling in data from Google spreadsheets.

But if you don't want to remember all that, there's rio.

4.1 What we'll cover

- Importing data from local or Web files into R
- Exporting data out of R
- Alternatives to rio

4.2 Packages needed in this chapter

- rio
- htmltab
- readxl
- googlesheets
- pacman
- janitor
- rmiscutils (GitHub) or readr
- tibble

4.3 The magic of rio

"The aim of rio is to make data file I/O [import/output] in R as easy as possible by implementing three simple functions in Swiss-army knife style," according to the project's GitHub page. Those functions are `import()`, `export()`, and `convert()`.

So, the rio package has just one function to read in many different types of files: `import()`. If you `import("myfile.csv")`, it knows to use a function to read a CSV file. `import("myspreadsheet.xlsx")` works the same way. In fact, rio handles more than two dozen formats including tab-separated data (with the extension .tsv), JSON, Stata, and fixed-width format data (.fwf).

Once you've analyzed your data, if you want to save the results as a CSV, Excel spreadsheet, or other format, rio's `export()` function can handle that.

You should have installed rio in Chapter 1, but if you skipped that part, install it now with `install.packages("rio")` .

I've set up some sample data with Boston winter snowfall data. You could head to http://bit.ly/BostonSnowfallCSV and right click to save the file as BostonWinterSnowfalls.csv in your current R project working directory. But one of the points of scripting is to replace manual work - tedious or otherwise - with automation that is easy to reproduce. Instead of clicking to download, you can use R's download.file function with the syntax `download.file("url", "destinationFileName.csv")`:

```
download.file("http://bit.ly/BostonSnowfallCSV", "BostonWinterSnowfalls.csv")
```

This assumes that your system will redirect from that bit.ly URL shortcut and successfully find the real file URL, https://raw.githubusercontent.com/smach/NICAR15data/master/BostonWinterSnowfalls.csv . I've occasionally had problems accessing Web content on old Windows machines. If you've got one of the those and this bit.ly link isn't working, you can swap in the actual URL for the bit.ly link. (You may want to use the shortcut link in your browser to get to the longer one that you can copy and paste, instead of typing it out manually. Or, go to http://bit.ly/R4JournalismLinks and copy the link from there). Another option is upgrading your Windows machine to Windows 10 if possible to see if that does the trick.

If you wish that rio could just import data directly from a URL, in fact it can, and I'll get to that in the next section. The point of *this* section is to get practice working with a local file.

Once you have the test file on your local system, you can load that data into an R object called snowdata with the code:

```
snowdata <- rio::import("BostonWinterSnowfalls.csv")
```

Note that it's possible rio will ask you to re-download the file in binary format, in which case you'll need to run download.file("http://bit.ly/BostonSnowfallCSV", "BostonWinterSnowfalls.csv", mode='wb'). I want to remind you again about RStudio's tab completion options if you haven't been using them. If you type rio:: and wait, you'll get a list of all available functions. Type snow and wait, and you should see the full name of your object as an option. Use your up and down arrow keys to move between auto-completion suggestions. Once the option you want is highlighted, hit the tab key (or Enter) and the full object or function name will be added to your script.

You should see the object snowdata appear in your environment tab in the RStudio top right pane (if that top right pane is showing your command History instead of your Environment, select the Environment tab). snowdata should show that it has 76 "obs." – observations, or rows – and 2 variables, or columns. If you click on the arrow to the left of snowdata to expand the listing, you'll see the 2 column names and the type of data each column holds. The Winter is character strings and the Total column is numeric. You should also be able to see the first few values of each column in the Environment pane.

Click on the word snowdata itself in the Environment tab for a more spreadsheet-like view of your data, as shown in Figure 4.1. You can get that same view from the R console with the command `View(snowdata)` (that's got to be a capital V on View - view won't work). Note: snowdata is not in quotation marks because you are referring to the name of an *R object in your environment.* In the rio::import command before, "BostonWinterSnowfalls.csv" *is* in quotation marks because that's not an R object; it's a character string name of a file outside of R.

This view has a couple of spreadsheet-like behaviors. Click on a column header and it will sort by that column's values in ascending order; click the same column header a second time, and it will sort descending. There's a search box to find rows matching certain characters.

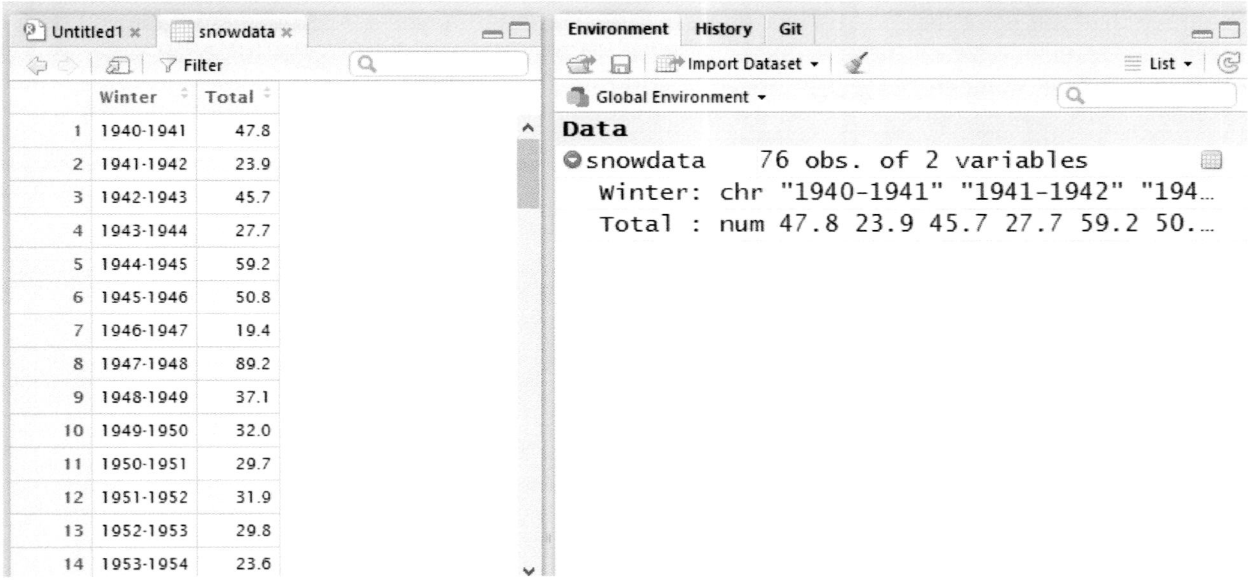
Untitled1
snowdata
Filter
Winter
Total
1 1940-1941 47.8
2 1941-1942 23.9
3 1942-1943 45.7
4 1943-1944 27.7
5 1944-1945 59.2
6 1945-1946 50.8
7 1946-1947 19.4
8 1947-1948 89.2
9 1948-1949 37.1
10 1949-1950 32.0
11 1950-1951 29.7
12 1951-1952 31.9
13 1952-1953 29.8
14 1953-1954 23.6
Environment
History
Git
Import Dataset
List
Global Environment
Data
snowdata 76 obs. of 2 variables
Winter: chr "1940-1941" "1941-1942" "194...
Total : num 47.8 23.9 45.7 27.7 59.2 50....

Figure 4.1: Viewing the snowdata R object within RStudio

If you click the Filter icon, you'll get a filter for each column. The Winter character column works as you might expect, filtering for any rows that contain the characters you type in. If you click in the Total numerical column's filter, though.

4.3.1 Import a file from the Web

If you want to download and import a file from the Web, you can do so if it's publicly available and in a format such as Excel or CSV. Try

```
snowdata <- rio::import("http://bit.ly/BostonSnowfallCSV", format = "csv")
```

A lot of systems will be able to follow the redirect URL to the file even after first giving you an error message, as long as you specify the format as "csv" since the file name here doesn't include ".csv". If yours won't, use the URL "https://raw.githubusercontent.com/smach/R4JournalismBook/master/data/BostonSnowfall.csv" instead.

rio can also import well-formatted HTML tables from Web pages, but the tables have to be *extremely* well-formatted. Let's say you want to download the table that describes the National Weather Service's severity ratings for snowstorms. The National Centers for Environmental Information Regional Snowfall Index page has just one table, very well crafted, so code like this should work:

```
rsi_description <- rio::import("https://www.ncdc.noaa.gov/snow-and-ice/rsi/", format="html")
```

Note again that you need to include the file format, in this case `format="html"`, because the URL itself doesn't give any indication as to what kind of file it is. If the URL included a file name with an .html extension, rio would know.

In real life, though, Web data rarely appears in such neat, isolated form. A good option for cases that aren't quite as well crafted is often the htmltab package. Install it with `install.packages("htmltab")`. The package's function for reading an HTML table is also called htmltab. But if you run this:

```
library(htmltab)
citytable <- htmltab("https://en.wikipedia.org/wiki/List_of_United_States_cities_by_population")
str(citytable)
```

you'll see that you don't have the correct table, because the data frame contains one object. Since I didn't specify *which* table, it pulled the first HTML table on the page. That didn't happen to be the one I want. I don't feel like importing every table on the page until I find the right one, but fortunately, I have a Chrome extension called Table Capture that lets me view a list of all the tables on a page, such as in Figure 4.2.

The last time I checked, table #5 with more than 300 rows was the one I wanted. If that doesn't work for you now, try installing Table Capture on a Chrome browser to check which table you want to download.

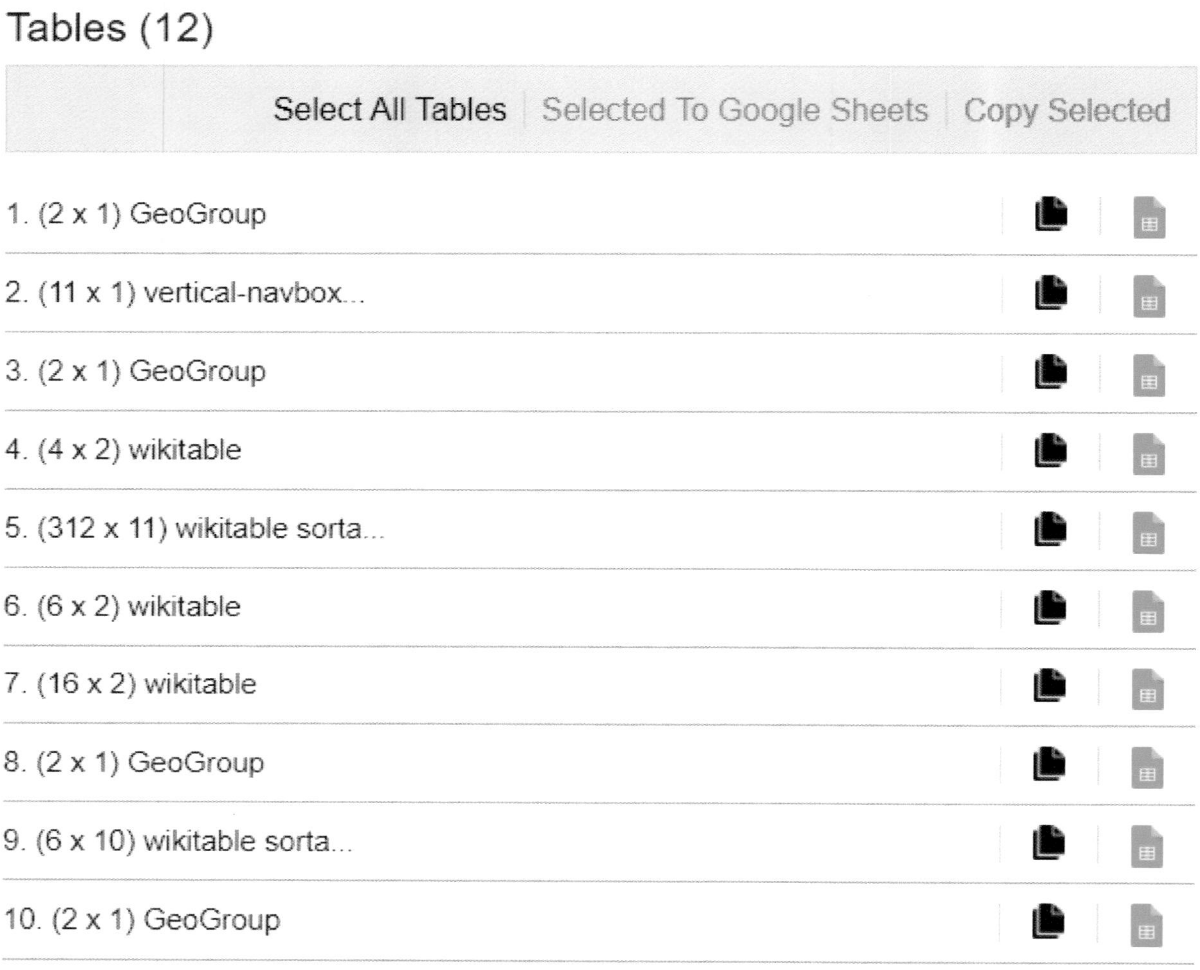

Tables (12)

Select All Tables | Selected To Google Sheets | Copy Selected

1. (2 x 1) GeoGroup
2. (11 x 1) vertical-navbox...
3. (2 x 1) GeoGroup
4. (4 x 2) wikitable
5. (312 x 11) wikitable sorta...
6. (6 x 2) wikitable
7. (16 x 2) wikitable
8. (2 x 1) GeoGroup
9. (6 x 10) wikitable sorta...
10. (2 x 1) GeoGroup

Figure 4.2: The Chrome Table Capture extension.

I'll try again, specifying table 5 and then seeing what column names are in the new citytable. Note that in the following code, I put the citytable <- htmltab() command onto multiple lines. That's so it didn't run off the printed page – you can keep everything on a single line. If the table number has changed since publication, replace which = 5 with the correct number.

Instead of using the page at Wikipedia, you can replace the Wikipedia URL with the URL of a copy of the file I created. That file is at http://bit.ly/WikiCityList. To use that version, type bit.ly/WikiCityList into a browser, then copy the lengthy URL it redirects to and use *that* instead of the wikipedia.org URL below:

```
library(htmltab)
citytable <- htmltab(
  "https://en.wikipedia.org/wiki/List_of_United_States_cities_by_population",
```

```
  which = 5)
colnames(citytable)

##  [1] "2017 rank"                "City"
##  [3] "State"                    "2017 estimate"
##  [5] "2010 Census"              "Change"
##  [7] "2016 land area"           "2016 land area"
##  [9] "2016 population density" "2016 population density"
## [11] "Location"
```

How did I know **which** was the argument I needed to specify the table number? I read the htmltab help file using the command **?htmltab**. That included all available arguments. I scanned the possibilities, and "**which** a vector of length one for identification of the table in the document" looked right.

Note, too, that I used colnames(citytable) instead of names(citytable) to see the column names. Either will work. Base R also has a rownames() function.

Anyway, those table results are a lot better, although we can see from running `str(citytable)` that a couple of columns which should be numbers came in as character strings. You can see this both by the chr next to the column name and quotation marks around values like "8,550,405".

This is one of R's small annoyances: R generally doesn't understand that 8,550 is a number. I dealt with this problem myself by writing my own function in my own rmiscutils package to turn all those "character strings" that are really numbers with commas back into numbers. Anyone can download the package from GitHub and use it.

The most popular way to install packages from GitHub is to use a package called devtools. devtools is an extremely powerful package designed mostly for people who want to write their *own* packages, and it includes a few ways to install packages from other places besides CRAN. However, devtools usually requires a couple of extra steps to install compared to a typical package, and I want to leave annoying system-admin tasks for a bit later.

However, the pacman package I suggested you install in Chapter 2 will also install packages from non-CRAN sources like GitHub. If you haven't yet, install pacman with `install.packages("pacman")`.

pacman's p_install_gh("username/packagerepo") function installs from a GitHub repo.

p_load_gh("username/packagerepo") _loads a package into memory if it already exists on your system and first installs then loads a package from GitHub if the package doesn't exist locally.

My rmisc utilities package can be found at "smach/rmiscutils". Run

```
pacman::p_load_gh("smach/rmiscutils")
```

and you'll install my rmiscutils package.

Note: An alternative package for installing packages from GitHub is called remotes, which you can install with:

```
install.packages("remotes")
```

Its main purpose is to install packages from remote repositories such as GitHub. You can look at the help file with `help(package="remotes")`.

And, possibly the slickest of all is a package called githubinstall. That aims to guess the repo where a package resides. Install it with

```
install.packages("githubinstall")
```

and then you can install my rmiscutils package using

```
githubinstall::gh_install_packages("rmiscutils")
```

You'll be asked if you want to install the package at smach/rmisutils.

Now that you've installed my collection of functions, you can use my number_with_commas() function to change those character strings that should be numbers back into numbers. I strongly suggest adding a new column to the data frame instead of modifying an existing column – that's good data analysis practice no matter what platform you're using.

I'll call the new column PopEst2017. (If the table has been updated since, use appropriate column names.)

```
library(rmiscutils)
citytable$PopEst2017 <- number_with_commas(citytable$`2017 estimate`)
```

My rmiscutils package isn't the only way to deal with imported numbers that have commas, by the way. After I created my rmiscutils package and its number_with_commas() function, the tidyverse readr package was born. readr also includes a function that turns character strings into numbers, parse_number().

After installing readr, you could generate numbers from the 2017 estimate column with readr:

```
citytable$PopEst2017 <- readr::parse_number(citytable$`2017 estimate`)
```

One advantage of readr::parse_number() is that you can define your own locale() to control things like encoding and decimal marks, which may be of interest to non-U.S.-based readers. Run `?parse_number` for more information.

Note: If you didn't use tab completion for the 2017 estimate column, you might have had a problem with that column name if it has a space in it at the time you are running this code. If you see my code above, you'll notice there are backwards single quote marks around the column name. That's because the existing name had a space in it, which you're not supposed to have in R. That column name has another problem: It starts with a number, also generally an R no-no. RStudio knows this, and automatically adds the needed back quotes around the name with tab autocomplete

Bonus tip: There's an R package (of course there is!) called janitor that can automatically fix troublesome column names imported from a non-R-friendly data source. Install it with `install.packages("janitor")`. Then, you can create new clean column names using janitor's clean_names() function.

I'll create an entirely new data frame instead of altering column names on my original data frame, and run janitor's clean_names() on the original data. Then, check the data frame column names with names():

```
citytable_cleaned <- janitor::clean_names(citytable)
names(citytable_cleaned)
```

```
##  [1] "x2017_rank"                "city"
##  [3] "state"                     "x2017_estimate"
##  [5] "x2010_census"              "change"
##  [7] "x2016_land_area"           "x2016_land_area_2"
##  [9] "x2016_population_density"  "x2016_population_density_2"
## [11] "location"                  "pop_est2017"
```

You'll see the spaces have been changed to underscores, which are legal in R variable names (as are periods). And, all column names that used to start with a number now have an x at the beginning.

If you don't want to waste memory by having two copies of essentially the same data, you can remove an R object from your working session with the rm() function: `rm(citytable)`.

4.4 Import data from packages

In the last chapter, I showed how to import government and financial data directly into R using the quantmod package. There are other packages that will let you access data directly from R.

The aptly named weatherdata package on CRAN can pull data from the Weather Underground API, which has information for many countries around the world. (An API, or Application Programming Interface, is a service that makes data available in a structured way so developers can easily access it when designing their software.)

The rnoaa package, a project from the rOpenSci group, taps into several different U.S. National Oceanic and Atmospheric Administration data sets including daily climate, buoy, and storm information.

If you are interested in state or local government data in the US or Canada, you may want to check out RSocrata to see if an agency you're interested in posts data there. I've yet to find a complete list of all available Socrata data sets, but there's a search page at https://www.opendatanetwork.com. Be careful though: There are community-uploaded sets along with official government data, so check a data set's owner and upload source before relying on it for more than R practice. "ODN Dataset" in a result means it's a file uploaded by someone in the general public. Official government data sets tend to live at URLs like https://data.CityOrStateName.gov or https://data.CityOrStateName.us. I'll walk you through pulling some New York City restaurant inspection data with RSocrata a few chapters from now.

For more data-import packages, see my searchable chart at http://bit.ly/RDataPkgs. Journalists and others in mass communication might be particularly interested in censusapi and tidycensus, both of which tap into U.S. Census Bureau data. Other useful government data packages include eu.us.opendata from the U.S. and European Union governments to make it easier to compare data in both regions, and cancensus for Canadian census data.

4.4.1 When the data's not ideally formatted

In all these sample data cases, the data has been not only well-formatted, but ideal: Once we found it, it was perfectly structured for R. What do I mean by that? It was rectangular, with each cell having a single value instead of merged cells. And the first row had column headers, as opposed to, say, a title row in large font across multiple cells in order to look pretty – or no column headers at all.

Dealing with "untidy" data can, unfortunately, get pretty complicated. But there are a couple of common issues that are easy to fix.

Beginning rows that aren't part of the data. If you know that the first few rows of an Excel spreadsheeet don't have data you want, you can tell rio to skip one or more lines. The syntax is `rio::import("mySpreadsheet.xlsx", skip=3)` to exclude the first three rows. skip takes an integer.

There are no column names in the spreadsheet. The default import assumes the first row of your sheet is the column names. If your data *doesn't* have headers, the first row of your data may end up as your column headers. To avoid that, use `rio::import("mySpreadsheet.xlsx", col_names = FALSE)` and R will generate default headers of X0, X1, X2 and so on. Or, use a syntax such as `rio::import("mySpreadsheet.xlsx", col_names = c("City", "State", "Population"))` to set your own column names.

If there are multiple tabs in your spreadsheet, the which argument will override the default of reading in the first worksheet. rio::import("mySpreadsheet.xlsx", which = 2) reads in the second worksheet.

4.5 What's a data frame? And what can you do with one?

rio imports a spreadsheet or CSV file as an R *data frame*. How do you know whether you've got a data frame? In the case of snowdata, `class(snowdata)` returns the class, or type, of object it is. `str(snowdata)` also tells you the class and adds a bit more information. Much of the info you see with str() is similar to what we saw in the RStudio environment pane: snowdata has 76 observations (rows) and 2 variables (columns).

I mentioned in Chapter 1 that data frames are somewhat like spreadsheets. However, data frames are more structured. Each column in a data frame is an R *vector*, which means that **every item in a column has to be the same data type**. One column can be all numbers and another column can be all strings, but within the column, data has to be consistent.

If you've got a data frame column with the values 5, 7, 4, and "value to come," R will not simply be unhappy and give you an error. Instead, it will "coerce" all your values to be the same data type. Since "value to come" can't be turned into a number, 5, 7 and 4 will end up being turned into character strings of "5", "7", and "4". This isn't usually what you want, so it's important to be aware of what type of data is within each column. One stray character string value in a column of 1,000 numbers can turn the whole thing into characters. If you want numbers, make sure you have them!

R does have a ways of referring to missing data that won't screw up the rest of your columns: NA means not available.

Data frames are rectangular: Each row has to have the same number of entries (although some can be blank), and each column has to have the same number of items.

Excel spreadsheet columns are typically referred to by letters: Column A, Column B, etc. You can refer to a data frame column with its name, by using the syntax `dataFrameName$columnName`. So, if you type `snowdata$Total` and hit enter, you'll see all the values in the Total column, as in Figure 4.3. (That's why when you run the `str(snowdata)` command, there's a dollar sign before the name of each column.)

```
> snowdata$Total
 [1]  47.8  23.9  45.7  27.7  59.2  50.8  19.4  89.2  37.1
[10]  32.0  29.7  31.9  29.8  23.6  25.1  60.9  52.0  44.7
[19]  34.1  40.9  61.5  44.7  30.9  63.0  50.4  44.1  60.1
[28]  44.8  53.8  48.8  57.3  47.5  10.3  36.9  27.6  46.6
[37]  58.5  85.1  27.5  12.5  22.3  61.8  32.7  43.0  26.6
[46]  18.1  42.5  52.6  15.5  39.2  19.1  22.0  83.9  96.3
[55]  14.9 107.6  51.9  25.6  36.4  24.4  45.9  15.1  70.9
[64]  39.4  86.6  39.9  17.1  51.2  65.9  35.7  81.0   9.3
[73]  63.4  58.9 110.6  36.2
```

Figure 4.3: The Total column in the snowdata data frame.

A reminder that those bracketed numbers at the left of the listing aren't part of the data, they're just telling you what position each line of data starts with. [1] means that line starts with the first item in the vector, [10] the 10th, etc.

RStudio tab completion works with data frame column names as well as object and function names. This is pretty useful to make sure you don't misspell a column name and break your script – and it also saves typing if you've got long column names.

Type snowdata$ and wait, and you'll see a list of all the column names in snowdata.

There are several ways to slice and dice data frames, which I'll get to in the next chapter.

It's easy to **add a column to a data frame**. Currently, the Total column shows winter snowfall in inches. To add a column showing totals in Meters, you can use this format:

```
snowdata$Meters <- snowdata$Total * 0.0254
```

The name of the new column is on the left, and there's a formula on the right. In Excel, you might have used `=A2 * 0.0254` and then copied the formula down the column. With a script, you don't have to worry about whether you've applied the formula properly to all the values in the column.

Now look at your snowdata object in the Environment tab. It should have a third variable, Meters.

Because snowdata is a data frame, it has certain data-frame properties that you can access from the command line. `nrow(snowdata)` will give you the numbers of rows and `ncol(snowdata)` the number of columns. Yes, you can view this in the RStudio environment to see how many observations and variables there are, but there will probably be times when you'll want to know this as part of a script. `colnames(snowdata)` or `names(snowdata)` will give you the name of snowdata columns. `rownames(snowdata)` will give you any row names (if none were set, it will default to character strings of the row number such as "1", "2", "3", etc.).

As discussed briefly in Chapter 3, some of these special dataframe functions (technically called "methods") not only give you information, but let you change characteristics of the data frame. So, `names(snowdata)` tells you the column names in the data frame, but

```
names(snowdata) <- c("Winter", "SnowInches", "SnowMeters")
```

will *change* the column names in the data frame.

You probably won't need to know all available methods for a data frame object, but if you're curious, `methods(class=class(snowdata))` will display them. To find out more about any method, run the usual help query with a question mark, such as `?merge` or `?subset`.

4.5.1 When a number's not really a number

Zip codes are a good example of "numbers" that shouldn't really be treated as such. Although technically "numeric," it doesn't make sense to do things like add two Zip codes together or take an average of Zip codes in a community. If you import a Zip-code column, though, R will likely turn it into a column of numbers. And if you're dealing with areas in New England where Zip codes start with 0, the 0 will disappear.

I have a tab-delineated file of Boston Zip codes by neighborhood, downloaded from a Massachusetts government agency, at https://raw.githubusercontent.com/smach/R4JournalismBook/master/data/bostonzips.txt. If I tried to import it with `zips <- rio::import("bostonzips.txt")`, the Zip codes come in as 2118, 2119, etc. and not 02118, 02119, and so on.

This is where it helps to know a little bit about the underlying function that rio's import() function uses. You can find those underlying functions by reading the import help file at `?import` .For pulling in tab-separated files, import uses either fread() from the data.table package or base R's read.table() function. The `?read.table` help says that you can specify column classes with the `colClasses` argument.

Create a data subdirectory in your current project directory, then download the bostonzips.txt file with `download.file("https://raw.githubusercontent.com/smach/R4JournalismBook/master/data/bostonzips.txt", "data/bostonzips.txt")`. If you import this file specifying both columns as character strings, the Zip codes will come in properly formated:

```
zips <- rio::import("data/bostonzips.txt", colClasses = c("character", "character"))
str(zips)
```

```
## 'data.frame':    35 obs. of  2 variables:
##  $ Zipcode      : chr  "02118" "02119" "02120" "02130" ...
##  $ Neighborhood: chr  "Boston South End" "Roxbury" "Roxbury Mission Hill" "Jamaica
## Plain" ...
```

Note that the column classes have to be set using the c() function, `c("character", "character")`. If you tried `colClasses = "character", "character"` , you'd get an error message. This is a typical error for R beginners, but it shouldn't take long to get into the c() habit.

A save-yourself-some-typing tip: Writing out c("character", "character") isn't all that arduous; but if you've got a spreadsheet with 16 columns where the first 14 need to be character strings, this can get annoying. R's rep() function can help. rep(), as you might have guessed, repeats whatever item you give it however many times you tell it to, using the format `rep(myitem, numtimes)`. `rep("character", 2)` is the same as c("character", "character"), so `colClasses = rep("character", 2)` is equivalent to `colClasses = c("character", "character")` .And, colClasses = c(rep("character", 14), rep("numeric", 2)) would set the first 14 columns as character strings and the last two as numbers. All the names of column classes here need to be in quotations marks because names are character strings.

I suggest you play around a little with rep() so you get used to the format, since it's a syntax that other R functions will use, too.

4.6 Easy sample data

R comes with some built-in data sets that are easy to use if you want to play around with new functions or other programming techniques. They're also used a lot by people teaching R, since instructors can be sure that all students are starting off with the same data in exactly the same format.

Type `data()` to see available built-in data sets in base R and whatever installed packages are currently loaded. `data(package = .packages(all.available = TRUE))` from base R will display all possible data sets from packages that are installed in your system, whether or not they're loaded into memory in your current working session.

You can get more information about a data set the same way you get help with functions: `?datasetname` or `help("datasetname")`. mtcars and iris are among those I've seen used very often.

If you type `mtcars`, the entire mtcars data set will print out in your console. You can use the head() function to look at the first few rows (more on head() in the next chapter) with `head(mtcars)`.

You can store that data set in another variable if you want, with a format like:

```
cardata <- mtcars
```

Or, running the data function with the data set name, such as `data(mtcars)`, loads the data set into your working environment.

One of the most interesting packages with sample data sets for journalists is the fivethirtyeight package, which has data from stories published on the FiveThirtyEight.com website. The package was created by several academics in consultation with FiveThirtyEight editors, and is designed to be a resource for teaching undergraduate statistics.

Pre-packaged data can be useful - and in some cases fun. In the real world, though, you may not be using data that's quite so conveniently packaged.

4.6.1 Create a data frame manually within R

Chances are, you'll often be dealing with data that starts off outside of R and you import from a spreadsheet, CSV file, API, or other source. But sometimes you might just want to type a small amount of data directly into R, or otherwise create a data frame manually. So let's take a quick look at how that works.

R data frames are assembled column by column by default, not one *row* at a time. If you wanted to assemble a quick data frame of town election results, you could create a vector of candidate names, a second vector with their party affiliation, and then a vector of their vote totals:

```
candidates <- c("Smith", "Jones", "Write-ins", "Blanks")
party <- c("Democrat", "Republican", "", "")
votes <- c(15248, 16723, 230, 5234)
```

Remember not to use commas in your numbers, like you might do in Excel.

To create a data frame from those columns, use the data.frame() function and the syntax data `(column1, column2, column3)`.

```
myresults <- data.frame(candidates, party, votes)
```

Check its structure with str():

```
str(myresults)
```

```
## 'data.frame':    4 obs. of  3 variables:
##  $ candidates: Factor w/ 4 levels "Blanks","Jones",..: 3 2 4 1
##  $ party     : Factor w/ 3 levels "","Democrat",..: 2 3 1 1
##  $ votes     : num  15248 16723 230 5234
```

While the candidates and party *vectors* are characters, the candidates and party *data frame columns* have been turned into a class of R objects called factors. It's a bit too in-the-weeds at this point to delve into how factors are different from characters, except to say that 1) factors can be useful if you want to order items in a certain, non-alphabetical way for graphing and other purposes, such as Poor is less than Fair is less than Good is less than Excellent; and 2) factors can behave differently than you might expect at times. I'd recommend sticking with character strings unless you have a good reason to specifically want factors.

You can keep your character strings intact when creating data frames by adding the argument stringsAsFactors = FALSE:

```
myresults <- data.frame(candidates, party, votes, stringsAsFactors = FALSE)
str(myresults)
```

```
## 'data.frame':    4 obs. of  3 variables:
##  $ candidates: chr  "Smith" "Jones" "Write-ins" "Blanks"
##  $ party     : chr  "Democrat" "Republican" "" ""
##  $ votes     : num  15248 16723 230 5234
```

Now, the values are what you expected.

There's one more thing I need to warn you about when creating data frames this way. If one column is shorter than the other(s), R will sometimes repeat data from the shorter column - *whether or not you want that to happen.*

Say, for example, you created the election results columns for candidates and party but only entered votes results for Smith and Jones, not for Write-ins and Blanks. You might expect the data frame would show the other two entries as blank, *but you'd be wrong.* Try it and see, by creating a new votes vector with just two numbers, and using that new votes vector to create another data frame:

```
votes <- c(15248, 16723)
myresults2 <- data.frame(candidates, party, votes)
str(myresults2)
```

```
## 'data.frame':    4 obs. of  3 variables:
##  $ candidates: Factor w/ 4 levels "Blanks","Jones",..: 3 2 4 1
##  $ party     : Factor w/ 3 levels "","Democrat",..: 2 3 1 1
##  $ votes     : num  15248 16723 15248 16723
```

That's right, R re-used the first two numbers, which is definitely *not* what you'd want. If you try this with three numbers in the votes vector instead of two or four, R would throw an error. That's because each entry couldn't be recycled the same number of times.

If by now you're thinking, "Why can't I create data frames that don't change strings into factors automatically? And why do I have to worry about data frames re-using one column's data if I forget to complete all the data?" Hadley Wickham had the same thought. His tibble package creates an R class, also called tibble, that he says is a "modern take on data frames. They keep the features that have stood the test of time, and drop the features that used to be convenient but are now frustrating."

If this appeals to you, install the tibble package if it's not on your system and then try to create a tibble with

```
myresults3 <- tibble::tibble(candidates, party, votes)
```

and you'll get an error message that the votes column needs to be either 4 items long or 1 item long (tibble() will repeat a single item as many times as needed, but only for one).

Put the votes column back to 4 entries if you'd like to create a tibble with this data:

```
library(tibble)
votes <- c(15248, 16723, 230, 5234)
myresults3 <- tibble(candidates, party, votes)
str(myresults3)
```

```
## Classes 'tbl_df', 'tbl' and 'data.frame':    4 obs. of  3 variables:
##  $ candidates: chr  "Smith" "Jones" "Write-ins" "Blanks"
##  $ party     : chr  "Democrat" "Republican" "" ""
##  $ votes     : num  15248 16723 230 5234
```

It looks similar to a data frame – in fact, it *is* a data frame, but with some special behaviors, such as how it prints. You'll also notice that the candidates column is character strings, not factors.

If you like this behavior, go ahead and use tibbles. However, given how prevalent conventional data frames remain in R, it's still important to know about their default behaviors.

4.7 Exporting data

Often after you've wrangled your data in R, you'll want to save your results. Here are some of the ways to export your data that I tend to use most:

- Save to a CSV file with `rio::export(myObjectName, file="myFileName.csv")` and to an Excel file with `rio::export(myObjectName, file="myFileName.xlsx")`. rio understands what file format you want based on the extension of the file name. There are several other available formats, including .tsv for tab-separated data, .json for JSON and .xml for XML.
- Save to an R binary object that makes it easy to load back into R in future sessions. There are two options.

Generic `save()` will save one or more objects into a file, such as `save(objectName1, objectName2, file="myfilename.RData")`. To read this data back into R, you just use the command `load("myfilename.RData")` and all the objects return with the same names in the same state they had before.

You can also save a single object into a file with `saveRDS(myobject, file="filename.rds")`. The logical assumption would be that loadRDS would read the file back in, but instead the command is *readRDS* – and in this case, just the data has been stored, *not the object name.* So, you need to read the data into a new object name, such as `mydata <- readRDS("filename.rds")`.

There's a third way of saving an R object specifically for R: generating the R commands that would re-create the object instead of the object with final results. The base R functions for generating an R file to re-create an object are dput() or dump(). However, I find rio::export(myobject, "mysavedfile.R) even easier to remember.

Finally, there are additional ways to save files that optimize for readability, speed, or compression, which I mention in the Additional Resources section at the end of this chapter.

You can also export an R object into your Windows or Mac clipboard with rio: `rio::export(myObjectName, format = "clipboard")`. And, you can import data into R from your clipboard the same way:

`rio::import(file = "clipboard")`.

Bonus: rio's convert() function lets you - you guessed it - convert one file type to another without having to manually pull the data into and then out of R. See ?convert for more info.

Final point: RStudio lets you click to import a file, without having to write code at all. This isn't something I recommend until you're comfortable importing from the command line, since I think it's important to understand the code behind importing. But, I'll admit this can be a handy shortcut.

In the Files tab of RStudio's lower right pane, navigate to the file you want to import and click on it. You'll see an option to either View File or Import Dataset. Choose import, and you'll see a dialog that previews the data, allows you to modify how the data is imported, and previews the code that will be generated.

Make whatever changes you want and hit import, and your data will be pulled into R.

Next up: Some real-world data analysis at last.

4.8 Additional resources

rio alternatives. While rio is a great Swiss Army knife of file handling, there may be times when you want a bit more control over how your data is pulled into or saved out of R. In addition, there have been times when I've had a challenging data file that rio choked on but another package could handle. Some other functions and packages you may want to explore:

- Base R's read.csv() and read.table() to import text files (use `?read.csv` and `?read.table` to get more information). stringsAsFactors = FALSE is needed with these if you want to keep your character strings as character strings. write.csv() will save to CSV.
- rio uses Hadley Wickham's readxl package for reading Excel files. Another alternative for Excel is openxlsx, which can write to an Excel file as well as read one. Look at the openxlsx package vignettes for information about formatting your spreadsheets as you export.
- Wickham's readr package is also worth a look as part of the "tidyverse." readr includes functions to read CSV, tab-separated, fixed-width, Web logs, and several other types of files. readr prints out the type of data it has determined for each column – integer, character, double (non-whole numbers), etc. It creates tibbles.

Import directly from a Google spreadsheet. The googlesheets package lets you import data from a Google Sheet, even if it's private, by authenticating your Google account. The package is available on CRAN; install it with with `install.packages("googlesheets")`. After loading it with `library("googlesheets")`, read the excellent introductory vignette. At the time of this writing, the intro vignette was available within R at `vignette("basic-usage", package="googlesheets")`. If you don't see it, try `help(package="googlesheets")` and click on the "User guides, package vignettes and other documentation" link for available vignettes, or look at the package information on GitHub at https://github.com/jennybc/googlesheets.

'Scrape' data from Web pages with the rvest package and SelectorGadget browser extension or JavaScript bookmarklet. SelectorGadget helps you discover the CSS elements of data you want to copy that are on an HTML page; then rvest uses R to find and save that data. This is not a technique for raw beginners, but once you've got some R experience under your belt, you may want to come back and re-visit this. I have some instructions and a video on how to do this at http://bit.ly/Rscraping. RStudio has a webinar available on demand as well, at https://www.rstudio.com/resources/webinars/extracting-data-from-the-web-part-2/.

Alternatives to base R's save and read functions. If you are working with large data sets, speed may become important to you when saving and loading files. The data.table package has a speedy fread() function, but beware that resulting objects are data.tables and not plain data frames; some behaviors are different. If you want a conventional data frame, you can get one with the `as.data.frame(mydatatable)` syntax. The data.table package's fwrite() function is aimed at writing to a CSV file considerably faster than base R's write.csv().

Two other packages might be of interest for storing and retrieving data. The feather package saves in a binary format that can be read either into R or Python. And, the fst package's read.fst() and write.fst() offer fast saving and loading of R data frame objects – plus the option of file compression.

Chapter 5

Basic Data Exploration

5.1 Project: Weather data

Trying to tell a story about unusual weather? Whether you're covering a hot summer, snowy winter, or busy hurricane season, one of the first questions you'll want to answer is: How does this stack up with previous summers/winters/hurricane seasons? "We had 8.3 inches of rain last month" won't mean much to your audience unless they know if that's unusual.

In other words: **A data point alone isn't nearly as interesting as one in context.** So, when coming up with questions to ask your data, make sure at least one or two start off: "How does this compare with...?"

R can help you answer those questions. Let's take a look at some snowfall data and start doing some basic analysis.

5.2 What we'll cover

- Downloading data files you'll need for the rest of this book's projects
- Techniques for summarizing data sets, including `str()`, `summary()` and some functions in add-on packages
- Bracket notation for subsetting data
- The dplyr package for more elegant data slicing and dicing

5.3 Packages needed in this chapter

Installed but not loaded:

- Hmisc
- psych
- rio

Installed and loaded:

- pacman
- usethis
- dplyr
- skimr

If you know which packages you need to install, you can do so with base R's install.packages() function. For packages already installed, you can load them with base R's library(). However, I find the pacman package to be more convenient to set up packages for workshops and classes – as well as keeping my own multiple systems in sync.

The code below 1) installs the usethis, dplyr, and skimr packages onto your system if they're not already there and loads them into memory, and 2) installs Hmisc and psych onto your system, whether or not they're there, but doesn't load them into memory.

```
library(pacman)

p_load(usethis, dplyr, skimr)
p_install(Hmisc)
p_install(psych)
```

Skip the second and third lines if you've already got Hmisc and psych on your system and don't need to re-install.

5.4 Download this book's files

If you'd like to use R to download all the data files (and some R scripts) from this book's GitHub repository, run

```
usethis::use_course("https://github.com/smach/R4JournalismBook/archive/master.zip", "your/dir/path")
```

Replace "your/dir/path" with the local path on your system *under which* you'd like a new RStudio project created. On Windows, for example, you might use "C:/Users/YourName/Documents," and a new R4JournalismBook-master directory – and RStudio project – would be created inside your Documents directory.

If you *don't* want to create a new directory and project, you can download and unzip files manually. Head to the URL for this book's GitHub repository at https://github.com/smach/R4JournalismBook, and look for a green button that offers you the option to "Clone or download." See Figure 5.1. Click it, and you'll have an option to download the entire repository as a zipped file. You can then unzip and place files and the data subdirectory in an existing directory of your choice.

Figure 5.1: Clone or download choices

Note: If you've used git before and have it installed on your local system, you can use RStudio's File > New Project menu to select "Version Control - Checkout a project from a version control repository." Then, simply follow the prompts for a git project, using the book repo's URL: https://github.com/smach/R4JournalismBook.git. You may first have to tell RStudio the location of your local git installation. If so, first head to Tools > Global Options > Git/SVN.

Whichever way you choose, you should now have all the necessary data files in a data subdirectory on your local system.

5.5 Data summaries

During Boston's 2014-15 Snowpocalypse, a lot of us became endlessly fascinated by snowfall data. (Fun fact: Boston got more snow in one 23-day period than Chicago got during its snowiest-ever *entire winter.*) But while 110.6 inches certainly *sounds* like a lot of snow, it's hard to tell if that's truly unexpected without knowing about amounts in other winters.

So let's go back to snowfall data. Now, we'll look at data for three different cities, not just Boston. You should now have the *BostonChicagoNYCSnowfalls.csv* CSV file in your project's data subdirectory. If not, find it at http://bit.ly/BosChiNYCsnowfalls. Import it with

```
snowdata <- rio::import("data/BostonChicagoNYCSnowfalls.csv")
```

Don't forget RStudio autocomplete! If you type `data/Bost` between quotation marks and pause to see the autocomplete options, you shouldn't have to type the whole file name.

One of the first things worth doing after importing a data set is looking at the first few rows, the last few rows, and a summary of some basic stats. R has functions to help with all three. head() will show you the first six rows of the data frame and tail() shows you the last six:

```
head(snowdata)
```

```
##      Winter Boston Chicago  NYC
## 1 1940-1941   47.8    52.5 39.0
## 2 1941-1942   23.9    29.8 11.3
## 3 1942-1943   45.7    45.2 29.5
## 4 1943-1944   27.7    24.0 23.8
## 5 1944-1945   59.2    34.9 27.1
## 6 1945-1946   50.8    23.9 31.4
```

```
tail(snowdata)
```

```
##       Winter Boston Chicago  NYC
## 71 2010-2011   81.0    57.9 61.9
## 72 2011-2012    9.3    19.8  7.4
## 73 2012-2013   63.4    30.1 26.1
## 74 2013-2014   58.9    82.0 57.4
## 75 2014-2015  110.6    50.7 50.3
## 76 2015-2016   36.2    31.2 32.1
```

If you'd like to change that number of items, use the `n` argument. For example, `head(snowdata, n=10)` will show the first 10 rows.

This gives you a good initial feel of what the column structure looks like and also helps check if data got garbled toward the end of a file (particularly with spreadsheet data, when final rows can be total rows or footnotes). `str(snowdata)` will show you that it's a data frame of 76 rows and 4 columns – one character column and three numeric columns.

Aside: If you've got a tibble instead of a plain data frame and it has a lot of columns, `head()` will only display data *in columns that fit in your console window.* The display just *mentions* the other columns that don't fit. This is one area where Wickham and I disagree on desired default behavior. If this bugs you as well, you can turn the tibble back into a regular data frame by nesting as.data.frame() inside another function, such as `head(as.data.frame(snowdata))`.

Another difference between tibbles and base-R data frames: Typing the name of a data frame will print out the entire data set (at least until it reaches a maximum that's set in your R global options. You can see that maximum in RStudio under Tools > Global Options > Code > Display.) Wickham assumes that's not what you want, so typing the name of a tibble only shows the first 10 rows.

There are a few other functions besides str() and class() to give you basic structural information about a data frame. `dim(snowdata)` will show you the number of rows and columns, as well as `nrow(snowdata)` for the number of rows and `ncol(snowdata)` the number of columns. In addition to rownames() and colnames() (or just names() for column names), `dimnames(snowdata)` will print out both row and column names in an R *list*.

If you need to check whether a column is numeric, the `is.numeric()` function does this:

```
is.numeric(snowdata$Boston)
```

```
## [1] TRUE
```

Type `is.` in the RStudio console, and you'll see a dropdown list of similar functions such as is.data.frame(), is.character(), and is.vector().

If you need a *different* data type – such as the previous example of zip codes that make more sense as characters than numbers, there are also `as.` functions to convert data from one type to another. `as.character(20500)` will turn that number into a character string; `as.numeric("756")` will turn that into a number. The conversion has to be straightforward, though: `as.numeric("1,798")` won't work unless you get rid of the comma first.

5.6 Data 'interviews'

Data journalists sometimes talk about "interviewing" a data set, the way you might interview a human source for a story or broadcast report. Techniques are obviously different, but the challenge is the same: This source has a lot of interesting things he/she/it could tell you, *but you need to ask some probing questions.*

If you were interviewing a person, you'd probably do a little preparatory background research about your subject. It's not that different for a data set – you want to know some basics before diving in.

For a brief statistical summary of a data set, run the summary() function

```
summary(snowdata)
```

```
##     Winter              Boston          Chicago          NYC
##  Length:76          Min.   :  9.30   Min.   :14.30   Min.   : 2.80
##  Class :character   1st Qu.: 27.57   1st Qu.:29.30   1st Qu.:13.70
##  Mode  :character   Median : 42.75   Median :38.00   Median :24.25
##                     Mean   : 44.49   Mean   :40.88   Mean   :27.05
##                     3rd Qu.: 57.60   3rd Qu.:50.90   3rd Qu.:36.00
##                     Max.   :110.60   Max.   :89.70   Max.   :75.60
```

summary() isn't all that useful for the character column, but on the numeric columns, we see that the lowest snow total in Boston was 9.3 inches; the mean and median are close together, in the low to mid 40s; and 110.6 was the highest total.

If those few descriptive statistics seem a little thin to you, R has a lot of individual functions you can run on the `snowdata$Total` column to get more descriptive statistics: `sd(snowdata$Boston)` for standard deviation, `var(snowdata$Boston)` for the variance, and so on. You could then run those on the other 2 columns. Or, there are ways to run a function across multiple columns in a single command that we'll cover later.

But if you'd prefer to generate a more robust statistical summary than the `summary()` basics by using one command, a few R packages have created their own summary options.

The psych and Hmisc packages both have a describe() function – a great example of why you might want to use the `packagename::functionname()` syntax instead of loading both packages into memory, since otherwise you might not be sure which describe() you're using.

```
Hmisc::describe(snowdata)
```

Hmisc::describe() returns information about all the columns, character and numeric. For all of the columns, it calculates how many values are missing and also how many are distinct. For numeric columns, along with a few extra descriptive stats, it shows the lowest 5 and highest 5, which is fairly interesting for this data set.

psych::describe() is designed for numeric data only. If you try running it on the full data frame, it will throw an error. We'll run this later, after learning how to select only the numeric columns.

Finally, the skimr package's skim() function will show information on each column, including a little histogram for each numeric one, as in Figure 5.2.

```
skim(snowdata)
```

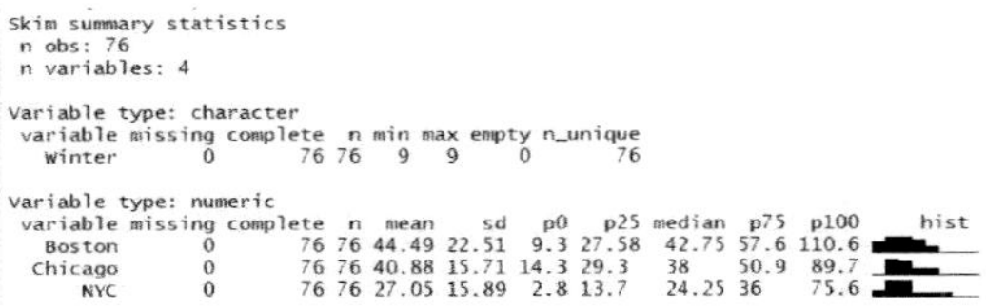

```
Skim summary statistics
 n obs: 76
 n variables: 4

Variable type: character
 variable missing complete  n min max empty n_unique
   Winter       0       76 76   9   9     0       76

Variable type: numeric
 variable missing complete  n  mean    sd   p0   p25 median  p75  p100
   Boston       0       76 76 44.49 22.51  9.3 27.58  42.75 57.6 110.6
  Chicago       0       76 76 40.88 15.71 14.3 29.3   38    50.9  89.7
      NYC       0       76 76 27.05 15.89  2.8 13.7   24.25 36    75.6
```

Figure 5.2: An example of skim() function output.

Next: some "interview questions".

5.7 Slicing and dicing your data set

Which winters had the most and least snow? How many winters had total snow of less than 2 feet or more than 6? Between 40 and 50 inches (close to an "average" Boston winter)? Data subsetting can answer basic questions like these.

This is one case where even for beginners, it's worth learning more than one way to do a task in R. Filtering is one of the most important skills to know in R, so I'd strongly suggest *not* skipping or lightly skimming this section, even if it gets a litte dry. Instead, grab a cup of your favorite caffeinated beverage, follow along carefully, and actually try out some of the code yourself.

Now, let's look at how to slice and dice.

Subset by some sort of condition. I'm going to show you two ways of doing this, and you can decide which one you like most.

I can tell you hands down which one I usually like the most: `dplyr`. I find Hadley Wickham's dplyr package generally to be a more elegant and human-readable way of subsetting data than *base R's bracket notation*. But even if you opt for dplyr, too, it's still important to understand R's bracket notation. Bracket notation can be used in a lot of different ways, including situations that dplyr might not cover.

So, top off your caffeinated beverage if needed, and try out some code.

1. The dplyr package. If I were stranded on an (Internet-limited) desert island and could only download one R package, this would be the one. Other journalists are fans as well; science reporter Peter Aldhous has called it a "game changing R package."

dplyr is exceptionally well thought out, and allows you to chain commands in a human-readable format. Let's take a look.

First, load the package with `library("dplyr")`. dplyr has its own head()-like function to examine the first few rows of a data frame, called `glimpse()` . It gives you a peek at a data frame similar to what you can see if you expand the data frame name in RStudio's top right Environment panel:

```
glimpse(snowdata)
```

dplyr also has several main "verbs" (creator Hadley Wickham's term) for dealing with data. The idea is that there are a few basic things we're all likely to want to do when wrangling data, and each one gets its own dplyr verb/function. These are the main tasks:

- Select certain rows based on a logical condition - dplyr's **filter()**. If you'd like to see all rows where Boston had more than 100 inches of snow: `filter(snowdata, Boston > 100)`. This says "filter the snowdata data frame to show only rows where the Boston column is greater than 100." (Spoiler: There were only two out of 76. 110 inches turns out to be a lot of snow for a Boston winter.)

To check for one condition OR another condition, use the `|` symbol, which means `or`. `filter(snowdata, Boston < 20 | Boston > 100)`, says "show rows in the snowdata data frame where the Boston column is less than 20 or greater than 100."

To filter by one condition AND a second condition, you can use the `&` sign, such as `filter(snowdata, Boston > 40 & Boston < 50)`. Actually, though, filter() *defaults* to "and" when there are multiple conditions. So, this code also works to find rows where Boston snow total is greater than 40 and less than 50: `filter(snowdata, Boston > 40, Boston < 50)`.

- Select certain rows based on row number - dplyr's **slice()**. It uses the format `myresults <- slice(snowdata, 60:76)`.

Note: In *all* these dplyr examples, the results need to be stored in a variable, such as `mynewdf <- slice(mydf, 6:10)` or `mydf <- slice(mydf, 6:10)`, That second code modifies the existing data frame; the first one creates a new data frame. To save space, I'm going to leave the `mydf <-` assignment part out of the rest of these examples, but remember that `slice(snowdata, 60:76)` will just print out the results in your console; it won't save them anywhere.

- Sort the data - dplyr's **arrange()**. Use `arrange(dataframe, colname)` to sort in ascending order and `arrange(dataframe, desc(colname))` to sort in descending order. Sort the snowdata data frame from Boston highest to lowest snowfalls with `arrange(snowdata, desc(Boston))`. To sort by a second column in case there are ties in the first column, the syntax is `arrange(dataframe, col1, col2)`.

- Select certain *columns* - dplyr's **select()**. You can select by specific column name, no quotes or c() needed: `select(snowdata, Winter, Boston)`. Select a contiguous group of columns, such as starting with Boston and ending with New York City, with the syntax `select(snowdata, Boston:NYC)`. You can select based on column names *containing certain characters*; for example, if you had a data frame with column names in the format `city_state` such as Boston_MA, Chicago_IL, NYC_NY, Fargo_ND and Syracuse_NY, you could select all the New York State entries using `select(dataframe, contains("_NY"))` or `select(dataframe, ends_with("_NY")`. You can delete columns by putting a minus sign before your selection, such as `select(snowdata, -(Boston:Chicago))` or `select(dataframe, -contains("_NY"))`.

`select_if()` lets you use `is.` functions such as `is.numeric()` or `is.character()` to choose columns by data type. If you want just the numeric columns in snowdata, for example: `select_if(snowdata, is.numeric)`. You could then run the psych package's describe() functions on those results, so only numeric columns remained:

```
snowdata_numeric <- select_if(snowdata, is.numeric)
psych::describe(snowdata_numeric)
```

Remember that `filter()` picks rows based on *data values*, but `select()` chooses columns based on the column *names* (not values of data within the columns).

dplyr also has functions for adding columns, creating summaries, splitting data into subgroups, merging two data frames, renaming columns, selecting highest and lowest by rank, and more. I'll cover more powers of dplyr throughout the book. For now, let's take a look at how to find some Chicago winters with low and high totals using dplyr's filter() and select(), and then sort from highest to lowest totals.

In the code below, the first row creates a temporary variable with all columns in the data set, but filtered just for rows where Chicago snow totals are less than 2 feet (24 inches) or more than 5 feet (60 inches). The second row takes that temporary data frame and selects the Winter and Chicago columns; the third line sorts the chicagoextremes data frame by snow totals from highest to lowest, storing it back into the same data frame.

```
chicagoextremes_allcolumns <- filter(snowdata, Chicago < 24 | Chicago > 60)
chicagoextremes <- select(chicagoextremes_allcolumns, Winter, Chicago)
chicagoextremes <- arrange(chicagoextremes, desc(Chicago))
```

Having to copy the data into a new data frame each time you want to do something wastes memory and typing time.

But dplyr has a very elegant way to fix this. It uses the %>% "pipe" symbol *that allows you to "pipe" the results of one function into a second function, without having to repeat the data frame name or store data in a temporary variable.* This automatically lets you break down complex operations into simple, more manageable chunks. It also makes code a lot more readable.

Here's the Chicago example, piped:

```
chicagoextremes <- snowdata %>%
  filter(Chicago < 24 | Chicago > 60) %>%
  select(Winter, Chicago) %>%
  arrange(desc(Chicago))
```

Piping is one of the most useful things you can learn in R, so it's worth taking a detailed look at how this works. What this code above is saying is:

Line 1: The new variable chicagoextremes is assigned all the data in the snowdata data frame. dplyr then "pipes" that result into line 2.

Line 2: The chicagoextremes data frame is filtered, keeping only rows where Chicago snowfall is less than 24 or greater than 60. dplyr then pipes *that* result to line 3.

Line 3: dplyr takes those results and then keeps only the Winter and Chicago columns. Since there's a %>% pipe at the end of this line also, dplyr sends this result to the next line.

Line 4: dplyr takes the results from line 4 and sorts the data frame by the Chicago column in descending order.

dplyr is incredibly powerful. If you've got, say, a spreadsheet of political donations by candidate, political party, and Zip code, you can use dplyr to easily group the data by zip code and party to find average and median donations for each. I'll be covering dplyr's `group_by()` function in Chapter 8.

2. Bracket notation: subsetting by row and column *numbers*. This is fairly simple subsetting, although you need to know the row and column numbers you want. If you happen to know that you want rows 60 to 76 in the second column of a data frame, you can store those in a new variable with `newdata <- snowdata[60:76,2]`. For **bracket notation** more generally, the syntax is `mydata[rownums, colnums]`.

You can select the single value in the first row and second column with `newdata <- snowdata[1,2]`. Once again, it's row number first, then column number. If you want *all* the columns, leave the column portion after the comma blank. So, selecting the entire first row would be `newdata <- snowdata[1,]`.

To choose *all* the rows in the second *column*, leave the row portion before the comma blank, such as `newdata <- snowdata[,2]`.

How would you select the second and third rows and the 3rd and 4th columns? `newdata <- mydata[2:3,3:4]`.

To create a new data frame for Boston and New York data excluding the Chicago column, you'd want (non-sequential) columns 1, 2 and 4. You can select non-adjacent columns with the c() function: `newdata <- snowdata[,c(1,2,4)]`. In this case, it might be easier to just leave out column 3, which you do with a minus sign: `newdata <- snowdata[,-3]`. To delete multiple columns or rows, use the minus sign before the -c() function, such as `newdata <- snowdata[,-c(2,4)]` for removing the Boston and NYC columns.

You can also select columns by *names*, not just numbers. Unlike with dplyr's select() function, in base R bracket notation, you need to put the column names in quotation marks.

```
snowdata[,c("Boston", "Chicago", "NYC")]
```

Bracket notation works for other types of R data, not just data frames. `snowdata$Boston[6]` will give you the 6th item in the snowdata Boston column, snowdata$Boston[-1] returns the full vector except for the first item, and snowdata$Boston[6:10] will give you items 6 through 10. (Because vectors don't have rows and columns, there's no need for a comma such as [6:10,] like you need for a data frame.)

Tip: If you're trying out these slightly different commands in your own console, remember that you can pull up the previous line of code you ran by using the up arrow. Then, you can edit that line of code – for example, changing [6] to [6:10] – instead of typing the whole command again. Or, you can type commands in the top-left script window, hit control/command enter to run them, and then see the results in the console.

If you want the last row of the data frame but don't want to use the actual row number 'snowdata[76,], what else could you do?

Well, that last row, 76, equals the *total number of rows in the data frame.* Hopefully, you remember that you can get the number of rows in a data frame with `nrow(snowdata)`. That means you can swap in `nrow(snowdata)` for a hard-coded row number. So in this case, `snowdata[76,]` is the same as `snowdata[nrow(snowdata),]`.

If you find these nested functions somewhat unreadable, one option is to break that single line of code into two: Store the number of rows in a *variable,* and then use that variable to swap in for 76:

```
lastrow <- nrow(snowdata)
snowdata[lastrow,]

##       Winter Boston Chicago  NYC
## 76 2015-2016   36.2    31.2 32.1
```

If you want to be sure that one line of code gives you the *same exact result* as another – same values *and* same structure – you can do a test with R's `identical()` function.

```
identical(snowdata[76,], snowdata[lastrow,])

## [1] TRUE

identical(snowdata[75,], snowdata[lastrow - 1,])

## [1] TRUE

identical(snowdata[75,], snowdata[lastrow,])

## [1] FALSE
```

The first two I compared *do* have identical results while the third pair does not (since the last row is 76, not 75).

As the second line of code above shows, it is possible to do calculations such as `lastrow - 1` inside the brackets.

What if you want to pull a row by a specific Winter, such as the 2014-2015 Snowpocalypse? You can use snowdata[snowdata$Winter == "2014-2015",].

Did you notice the *two* equals signs in snowdata$Winter == "2014-2015? Reminder, that's not a printing error; it's a **very important point:** When you're testing whether one thing is equal to another thing, **you have to use == and not =.** It would be the same for dplyr: `filter(snowdata, Winter == "2014-2015")`

If you've programmed in another computer language, this probably isn't news to you. If you're new to coding, though, be warned that this can trip up beginning programmers. Imprint this in your mind: =

performs an action. `myvariable = 5` *sets the value of* ***myvariable*** *to 5.* `==` *tests whether something is true:* `myvariable == 5` will tell you TRUE if myvariable equals 5 and FALSE if myvariable equals something else. (If myvariable doesn't exist, you'll get an error message).

We saw from the summary() and describe() functions what the lowest and highest snow totals were for each city. But what if we want to find the rows with those totals so we see can which winters they were?

R has functions to find minimum and maximum values: `min()` and `max()`. `max(snowdata$Boston)` will return the amount of snow during the Snowpocalypse, while `min(snowdata$Boston)` gets the amount of snow in Boston's least snowiest winter. And, `range(snowdata$Boston)` will return both the lowest and highest values.

Can you think of how you'd pull the row with the most Boston snow? You'd want it to be equal to the highest snow total, which is max(snowdata$Boston). With dplyr, it's `filter(snowdata, Boston == max(Boston))`. With bracket notation, it would be `snowdata[snowdata$Boston == max(snowdata$Boston),]`. I think the dplyr version is much more readable.

For finding most and least values, R also has functions `which.max()` and `which.min()` that give the *location* of the maximum and minimum values. `which.min(snowdata$Boston)` gives the index location of the lowest value (in this case number 72). So this would also work to pull the row from snowdata that has the lowest Boston winter snow total:

```
slice(snowdata, which.min(Boston))
```

This is how you could find Chicago winters with less than 2 feet of snow or more than 5 feet using bracket notation (with the second line sorting the results):

```
chicagoextremes <- snowdata[snowdata$Chicago < 24 | snowdata$Chicago > 60,
c("Winter", "Chicago")]
chicagoextremes <- chicagoextremes[order(-chicagoextremes$Chicago),]
```

To read the first line, you have to work from the inside out, while dplyr lets you see each operation sequentially on its own line. (The dplyr code can in fact also be on one line, but the format I used earlier is fairly typical and much more readable.)

I'll mostly be using dplyr syntax from here on, because I think it's a lot easier to follow.

5.8 More subsetting with dplyr

Now, let's ask another question of the snow data: Which Boston winters had snowfall of *at least* 72 inches (6 feet)?

Again we'll use a *condition*, this time that the Boston column is greater or equal to 72. I'll store the results in a new variable that I'll call boston72 and then view the results:

```
boston72 <- filter(snowdata, Boston >=72)
boston72
```

```
##      Winter Boston Chicago  NYC
## 1 1947-1948   89.2    38.1 63.2
## 2 1977-1978   85.1    82.3 50.7
## 3 1992-1993   83.9    46.9 24.5
## 4 1993-1994   96.3    41.8 53.4
## 5 1995-1996  107.6    23.9 75.6
## 6 2004-2005   86.6    39.4 41.0
## 7 2010-2011   81.0    57.9 61.9
## 8 2014-2015  110.6    50.7 50.3
```

Including data for other cities doesn't make sense for this question, so, let's select just the Winter and Boston columns, and then sort the results from highest to lowest:

```
boston72 <- filter(snowdata, Boston >=72) %>%
  select(Winter, Boston) %>%
  arrange(desc(Boston))
```

5.9 Wrap-Up

Important tools and techniques covered this chapter:

- Using the pacman package as well as base R to install and load packages
- Downloading this book's GitHub repo to your local sytem with usethis::use_course()
- Using functions such as str(), head(), tail(), summary(), Hmisc::describe(), psych::describe(), and skimr::skim() to explore a data set
- Getting a basic idea of the dplyr package's power and intuitive syntax for wrangling a data frame, including filter(), select(), and arrange()
- Seeing how base R's bracket notation accomplishes some of those same tasks

Next up: We'll get started with data visualization in R.

5.10 Additional resources

The dataMaid package was designed to generate a printable PDF report for a data set. The report includes a statistical summary of a data frame by default, but it's also possible to add information such as where the data came from. Just be warned that generating PDFs within R can be complicated to set up, especially on Windows, requiring versions of TeX/LaTeX and pandoc to be on your local system. dataMaid's visualize() function doesn't require PDF-creating capabilities, though. Blog post with more dataMaid details: http://sandsynligvis.dk/articles/18/codebook.html. The conflicted package will warn you if you try to use a function where there's more than one function with the same name loaded in your working session. Find out more at https://github.com/r-lib/conflicted.

Chapter 6

Beginning data visualization

"Don't tell, *show.*"

That's classic advice from journalism professors, and it's pretty useful when you want to communicate information with data. Visualization is one of the most important ways to explore a data set, and it's an area where R really shines.

What have Boston winters been like in the 21st century compared to the 20th? How do New York, Chicago, and Boston stack up in snowfall? Do they tend to have low and high snowfalls in the same years? What do the top 10 snowfall winters look like for each city? Visualizations can help you get insight into these types of questions and more.

6.1 Project: More weather data

We'll continue looking at snowfall data from Boston, Chicago, and New York, using various visualizations to compare them.

6.2 What we'll cover: How to ...

- Answer questions about your data with a few common visualization types
- Look at two platforms for visualization data: base R graphics and ggplot2
- Easily rename columns with dplyr's rename() function
- Build ggplot2 graphs layer by layer
- Understand what 'tidy' data means
- Create your own RStudio code snippets so you don't have to remember complex R syntax
- Take a stab at a more customized, presentation-worthy bar chart
- Comment your code
- Customize "out-of-the-box" ggplot2 graphs

6.3 Packages needed in this chapter

```
pacman::p_load(ggplot2, dplyr, usethis, forcats)
```

6.4 Answer questions with graphics

There are several types of questions that basic visualizations can help answer.

`summary()` returns the highest and lowest points in your data, but not *how unusual* those points are. `Hmisc::describe()` gives more statistical information. This helps to answer questions like: What's a typical value? Where are a majority of the values clustered? How do the outliers compare to the overall data set?

But looking at those lists of numbers may not have the same impact as visualizing them. Box plots and histograms, two types of graphics that you probably won't see in your local newspaper, are good ways to start exploring these types of questions visually and get a better understanding of your data's *distribution.*

How has a data set *changed over time*? Line charts over time are one way. Scatter plots are another way to visually scan for trends. R also lets you easily add trends lines to your graphs.

How do *two data sets compare*? You can look at data distributions side by side with box plots, create grouped bar charts, calculate and visualize statistical correlations, and more.

Bar charts can also help you (and your audience) see some simple *comparisons and trends.*

With R's roots as a statistical language, all of these visualization tasks are built in. Plus, R's large user community has contributed some compelling alternatives.

6.5 Easy visualizations in 1 or 2 lines of code

As with almost every task in R, there are multiple tools for visualizing data. The core R language has a lot of visualization functions, and many R users do sophisticated work with them. In addition, there are external dataviz packages that offer entire frameworks for visualizing data. As I'm writing this, Hadley Wickham's ggplot2 package is probably the best known and most popular of these; lattice is another. (Note that I'm referring specifically to *static* visualizations that can be saved as an image file; there are other packages for *interactive Web graphics.*)

I mostly use ggplot2, but I still turn to base R graphs from time to time, especially to do simple "I've just received a new data set" exploration. For more complex graphs where I want a significant amount of customization, I almost always use ggplot2.

Both types of graphics can be customized with colors, text type, axis labels, and more. I won't cover customizing graphs in base R, since I'll be working mostly with ggplot2 to tweak these types of settings. But if you're interested, you can get a taste of what's possible with base R graphics by running `demo(topic="graphics")` in your R console. (To see *all* available demos and not just graphics, including those that come with add-on packages installed on your system, run `demo()`).

6.5.1 Base R graphics

Let's focus for now on Boston snowfall data. If you don't still have the snowdata object from Chapter 5, load it with `snowdata <- rio::import("data/BostonChicagoNYCSnowfalls.csv")`.

This code gets just Boston snow data information:

```
bostonsnow <- select(snowdata, Winter, Boston)
```

and the next line of code renames the second column to "TotalSnow" using base R's names() function

```
names(bostonsnow)[2] <- "TotalSnow"
```

You can rename columns within dplyr, though, using the rename() function and the syntax `rename(newcolname = oldcolname)`

```
bostonsnow <- select(snowdata, Winter, Boston) %>%
  rename(TotalSnow = Boston)
```

In fact, you can select and rename columns all in one line, since dplyr's select() function will understand that `newcolname = oldcolname` means you want to rename a column in the data frame you're working on:

```
bostonsnow <- select(snowdata, Winter, TotalSnow = Boston)
```

The bracket notation format for renaming a column by name is `names(bostonsnow)[names(bostonsnow) == "Boston"] <- "TotalSnow"`. Compare that to `rename(snowdata, TotalSnow = Boston)`, and I'm guessing you can see why so many R users like dplyr. From here on, I'll be largely sticking to dplyr syntax.

6.6 Some basic graphs

Base R's plot() function will try to guess what kind of visualization you want, but it doesn't work with character strings. This will work to create the visualization in Figure 6.1:

```
plot(bostonsnow$TotalSnow)
```

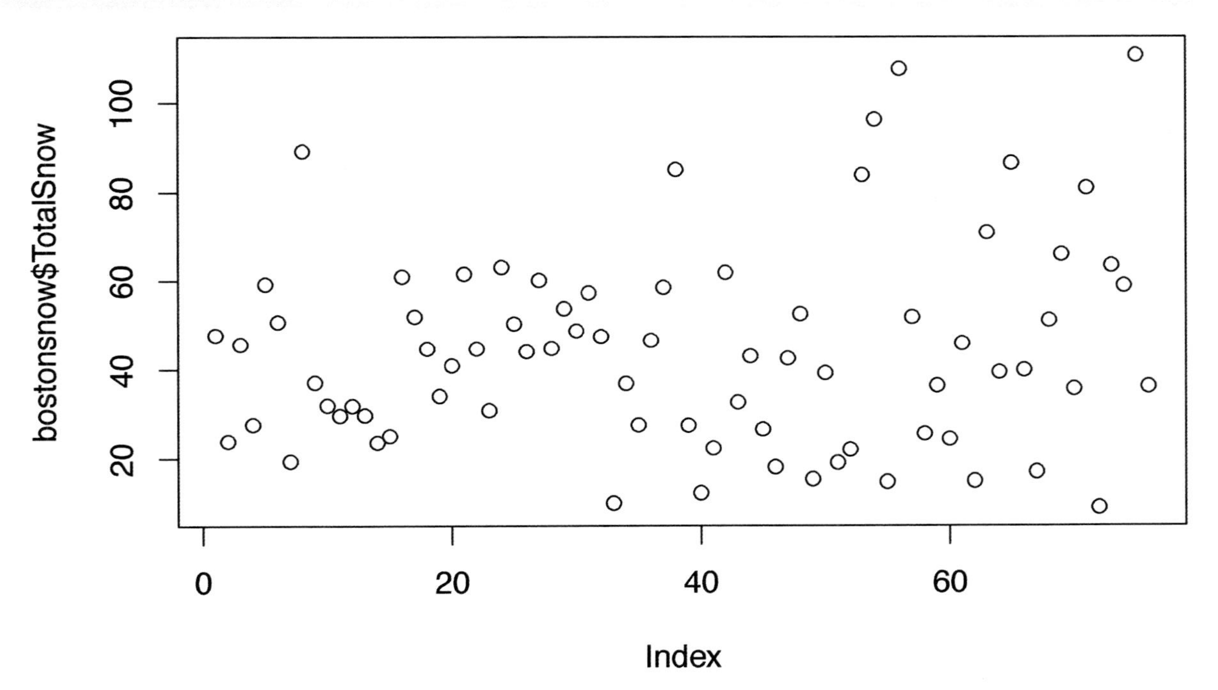

Figure 6.1: Base R's plot function.

because the data is numeric. But this won't:

```
plot(bostonsnow$Winter, bostonsnow$TotalSnow)
```

unless I first create a column of *factors* from that bostonsnow$Winter column of *character strings*.

The default `plot(bostonsnow$TotalSnow)`scatterplot gives some idea of the distribution of the snowfall totals, but not a very compelling one. So I'd like a different visualization.

You can pick which visualization type you want in base R, instead of allowing plot() to guess for you, by using a function that specifies the type. Choices include:

- **barplot()**

- **boxplot()**
- **dotchart()**
- **hist()** (histograms)
- **pie()** (piecharts)

Line charts don't have their own function, though. For line charts, the format is `plot(xvector, yvector, type = "l")`, and here, too, the X vector can't be character strings, although factors will work. I'll discuss factors in a bit more detail soon; but for now, if you're wondering how you might do that, simply wrapping the vector of character strings in factor() or as.factor() will work: `plot(factor(bostonsnow$Winter), bostonsnow$TotalSnow, type = "l")`.

Histograms are designed to show distributions, as in Figure 6.2, so let's look at a TotalSnow histogram:

```
hist(bostonsnow$TotalSnow)
```

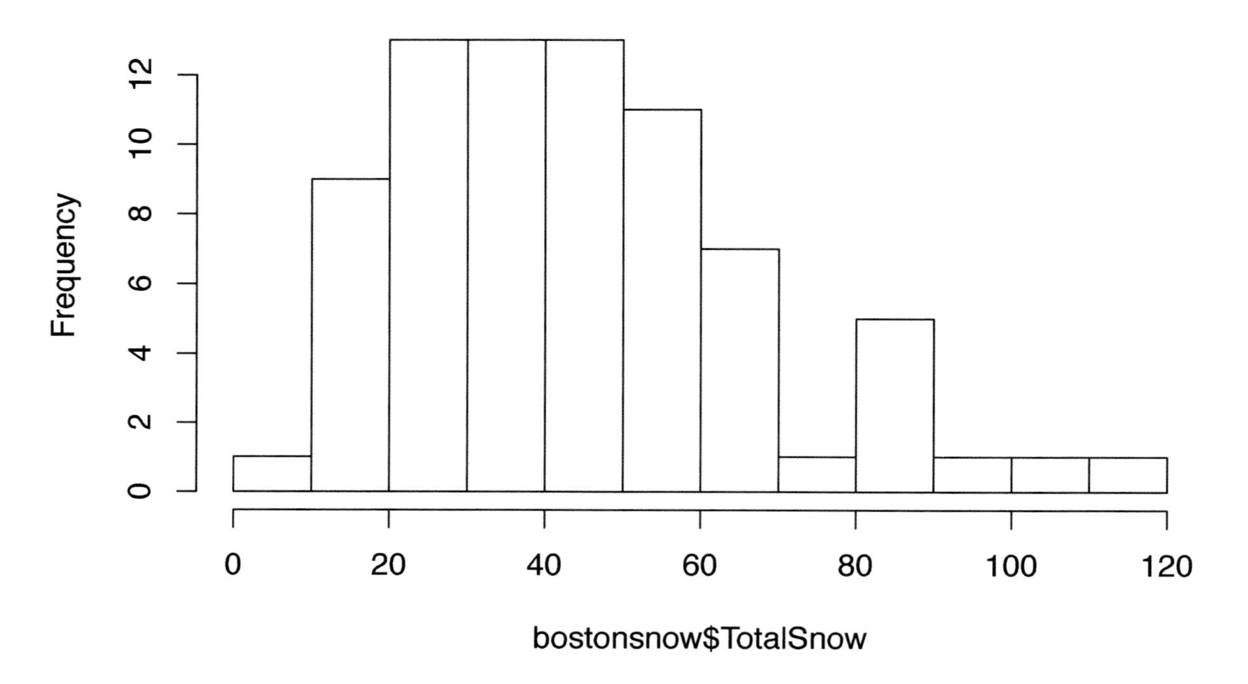

Figure 6.2: Base R histogram.

That gives a better idea of the distribution, with a fair number of winters between 20 and 50 inches and most between 10 and 70.

Here's a boxplot of the same data:

```
boxplot(bostonsnow$TotalSnow)
```

The "box"" in a boxplot such as Figure 6.3 shows where half the values fall. The bottom of the box represents what's called the "first quartile" – one quarter of all winters had snowfalls of less than that. The top of the box is the "third quartile" – three quarters of the winters had values less than that. The thick line in the box is the median: Half of the winters had more snow and half had less (or maybe slightly less than half if a bunch of winters had exactly the median amount). The area within the box between the first and third quartiles is known in statistics lingo as the "interquartile range."

The "whiskers" coming off the boxplot, those dotted lines that end with a short perpendicular solid line, aim

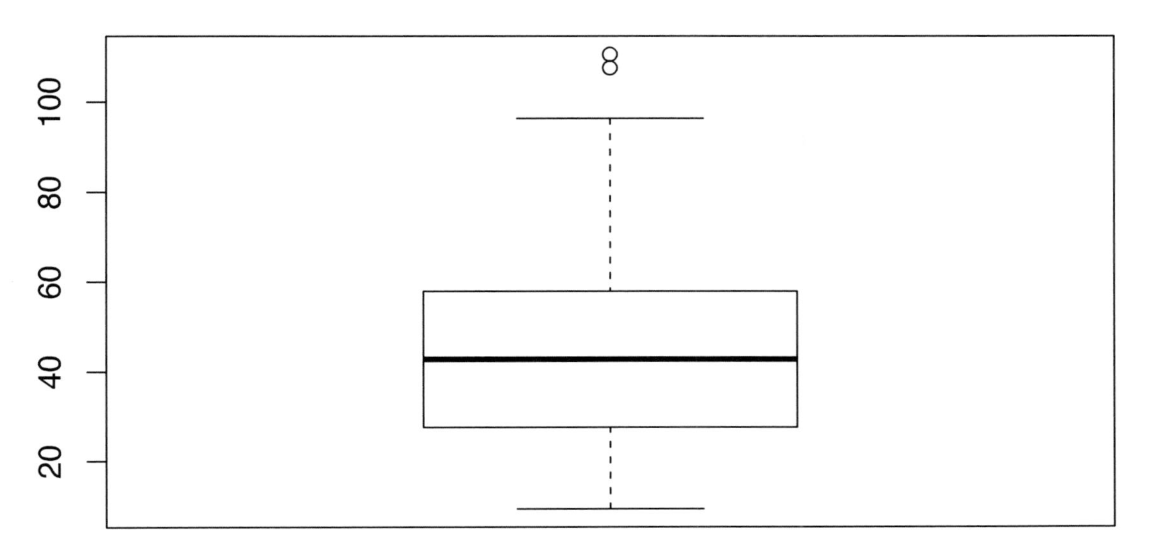

Figure 6.3: Base R boxplot.

to give a rough idea of where you might expect the rest of the values to fall. Base R's default is that the whiskers are 1.5 times higher and lower than the values in the interquartile range. Any points beyond the whiskers are considered outliers and are represented by small circles. We can see that the two winters above 100 inches of snow are statistical outliers.

That's one way of looking at the data. Another is a bar chart. For a very simple bar chart without sorting data or labeling the x axis, this will work:

```
barplot(bostonsnow$TotalSnow)
```

Visualizing winter snowfall amounts in proper time sequence can help detect any trends over time in the data, but it's not very useful if the goal is to look at high amounts in context. The base R function `sort()` can help here:

```
barplot(sort(bostonsnow$TotalSnow))
```

`sort(bostonsnow$TotalSnow, decreasing = TRUE)` will sort the data in descending order, so creating a bar chart with the bars in descending order would use the format:

```
barplot(sort(bostonsnow$TotalSnow, decreasing = TRUE))
```

How could you find out that to sort in descending order as in Figure 6.4 you need `decreasing = TRUE` syntax? Read the sort help file with `?sort`.

6.6.1 Quick plots with ggplot2

There are two main plotting platforms within ggplot2 itself. This may sound confusing, but the idea is to help those who are just starting out and not make things even more baffling. Package creator Hadley Wickham would probably like us all to use the full-blown, more powerful ggplot2 alternative, but he's enough of a realist to know that sometimes we users would just like to dash off a simple exploratory graphic, without worrying about multiple layers and half a dozen tweaks.

For that "I just want to do something simple *right now*" case, there's **qplot()** (short for quickplot). Like base R's plot(), ggplot's qplot() will try to guess an appropriate visualization type depending on the data you feed into it. For Boston snow, one option is:

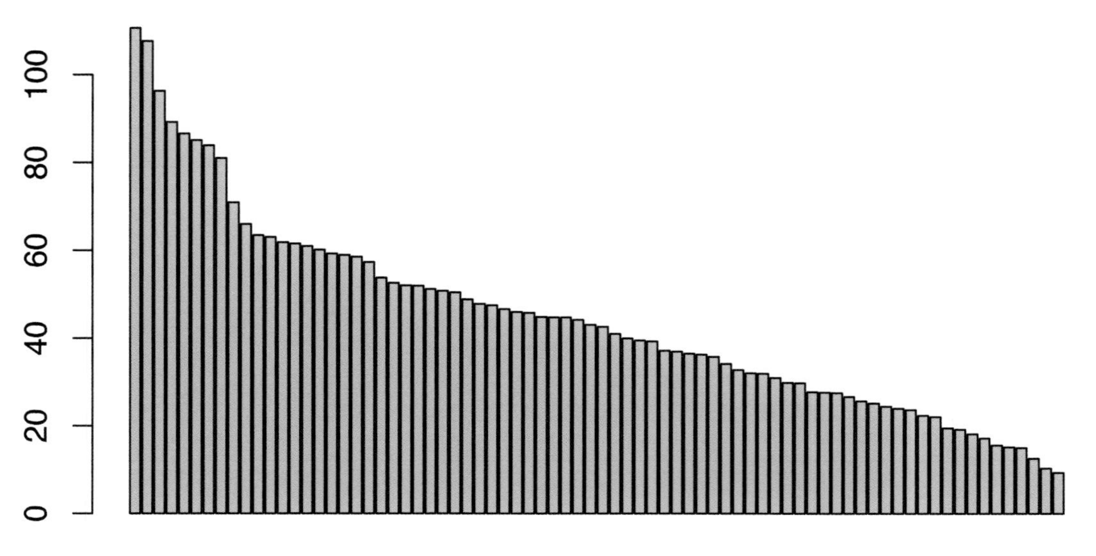

Figure 6.4: Sorted bar chart in base R.

```
qplot(data=bostonsnow, y = TotalSnow)
```

But ggplot2's plot()-like format also works to create a plot as in Figure 6.5, specifying the data for the y axis:

```
qplot(y = bostonsnow$TotalSnow)
```

However, I don't see a big advantage in learning qplot() instead of base R graphics. If you're going to use ggplot2, it's worth learning ...

6.7 The full power of ggplot2

It will take a little time and effort to fully understand how ggplot2 works, but the payback is a great deal of power and flexibility, as well as ease of re-using one complex graphic with different data sets. I won't claim that the syntax is *intuitive* – in my opinion, it's not nearly as user-friendly as dplyr – but it *is* fairly logical. At least there's a consistent structure, regardless of the type of visualization you want to create.

ggplot2 is based on a theory of visualization called the "grammar of graphics" (hence the gg). Using this "grammar" structure, you begin a visualization by creating a simple plot foundation, and then add various attributes, or layers, to bring it to life.

ggplot2 is the package name, but ggplot() (without the 2) is its key function.

A ggplot() visualization starts with data (obviously) and a couple of basic "aesthetics" (shortened to `aes` in code). I think of aesthetics as a data *concept* that, for example, assigns one data column to the X axis, another data column to the y axis, and maybe another column of data to change the color or size of items in the graph. But **the aesthetics layer by itself won't actually display anything.** This is an important point for understanding how to build a graph. This part of the code just sets up a structural foundation for a graphic, but it doesn't say whether the viz should be a bar chart, a line chart, a scatter plot, etc.

A first line of code for a ggplot() graphic looks something like `ggplot(mydf, aes(x=myxcolumn, y=myycolumn))`.

A second layer with geoms (geometric objects, remember them?) specifies the *type* of plot you want (line, bar, etc.). In full ggplot(), unlike qplot(), geoms are separate functions that start with `geom_` such as geom_bar(), geom_boxplot(), geom_histogram(), geom_point(), and so on.

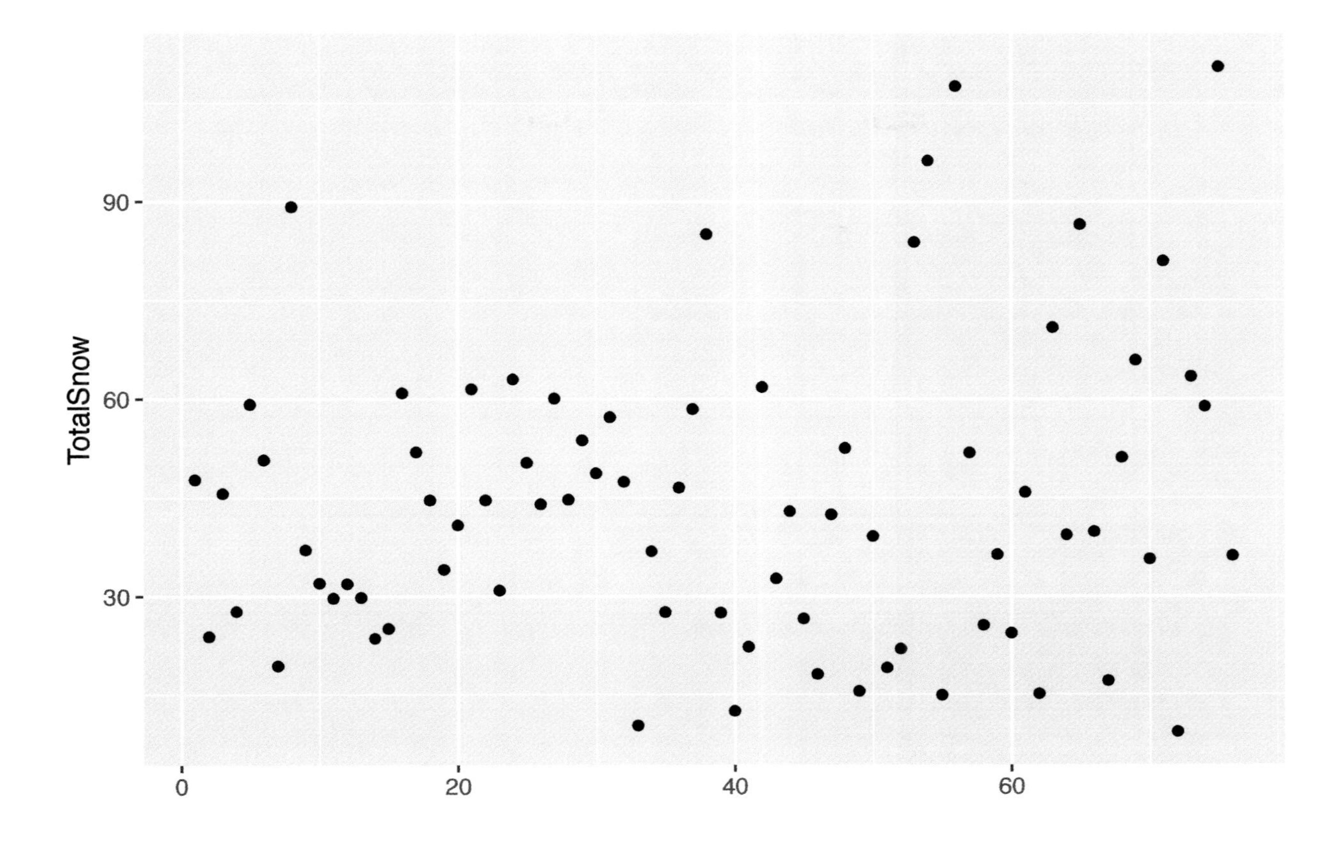

Figure 6.5: Using ggplot2's qplot() function.

Building on the foundational first line of code might therefore look something like:

```
ggplot(mydf, aes(x = myxcolumn, y = myycolumn)) +
  geom_point()
```

You need at least these two layers, ggplot() and a geom_ function, in order to display anything. And unlike with dplyr, additional lines of ggplot2 code are added with a `+` sign, not a `%>%` pipe.

Because all the geom functions start with geom_, typing geom_ in RStudio gives you an autocomplete drop-down list of available geom_functions.

From here, you can add a few or a *lot* of other layers to customize a visualization. Some layers can change your axes, modify labels, add a title, set a color palette, and more.

6.8 Basic ggplot2 customizations

Remember the qplot() box plot from the last section? The code for this looks pretty similar in full ggplot(), creating the visualization in Figure 6.6:

```
ggplot(data=snowdata) +
  geom_boxplot(aes(x = "Boston", y = Boston))
```

However, the difference is: *I can keep adding more layers.*

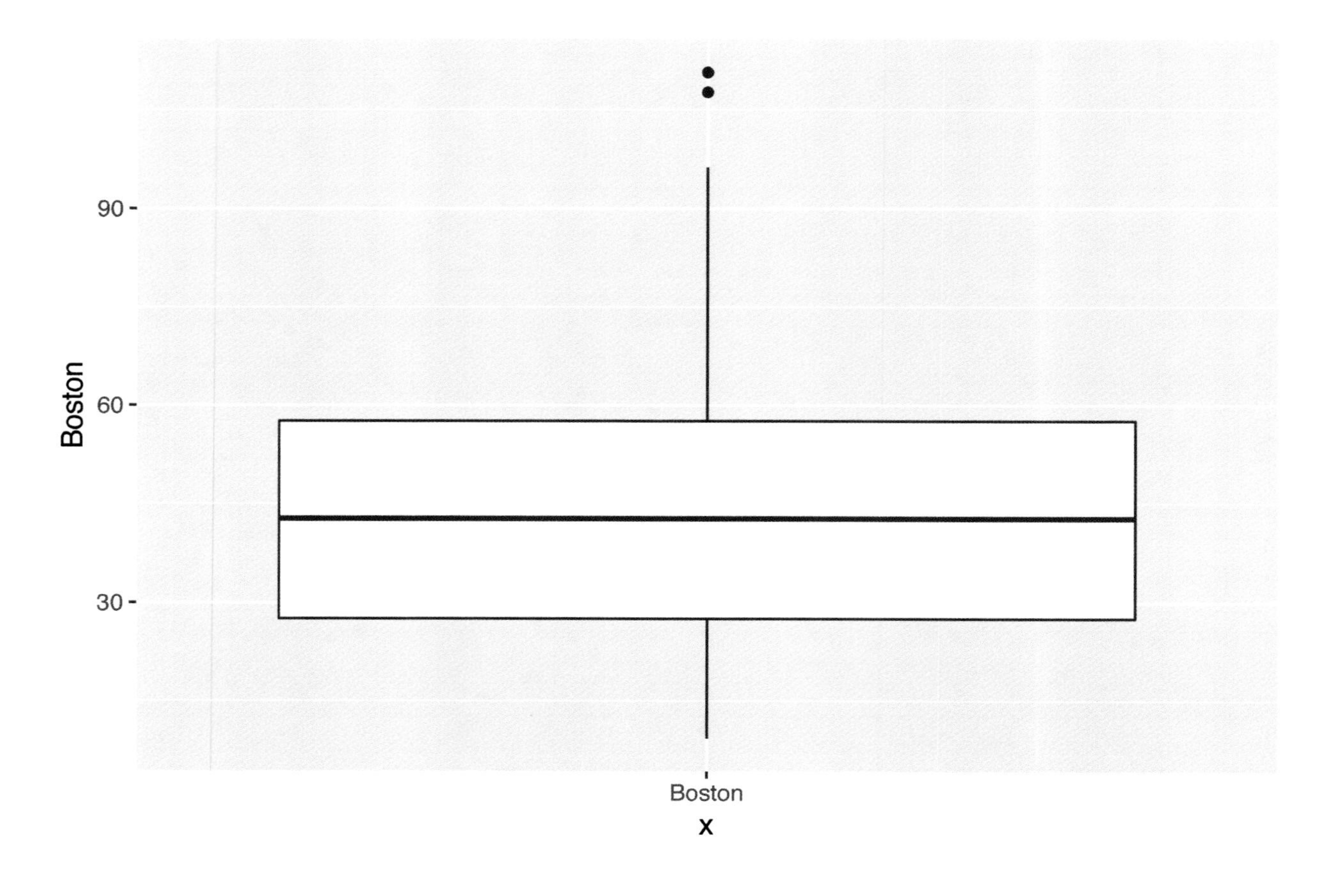

Figure 6.6: Box plot created with the ggplot() function.

So now if I'd like to compare that Boston snowfall distribution to Chicago's distribution, I just add another geom_boxplot() layer to produce a plot as in Figure 6.7:

```
ggplot(data=snowdata) +
  geom_boxplot(aes(x = "Boston", y = Boston)) +
  geom_boxplot(aes(x = "Chicago", y = Chicago))
```

Try adding NYC the same way. You should have a plot similar to Figure 6.8:

This works just fine if you've got a couple of columns of data, but not necessarily if you've got 5 or 6 – or if you may be adding columns to the data set later. Or, if you need to rename one of the categories. There's a more elegant way to handle this, but it involves having the data in another, "tidy" format.

As a reminder, this is the current data structure:

```
head(snowdata)
```

```
##      Winter Boston Chicago  NYC
## 1 1940-1941   47.8    52.5 39.0
## 2 1941-1942   23.9    29.8 11.3
## 3 1942-1943   45.7    45.2 29.5
## 4 1943-1944   27.7    24.0 23.8
## 5 1944-1945   59.2    34.9 27.1
## 6 1945-1946   50.8    23.9 31.4
```

A *tidy* verion of that would only have one observation in each row. Above, we've got three different data

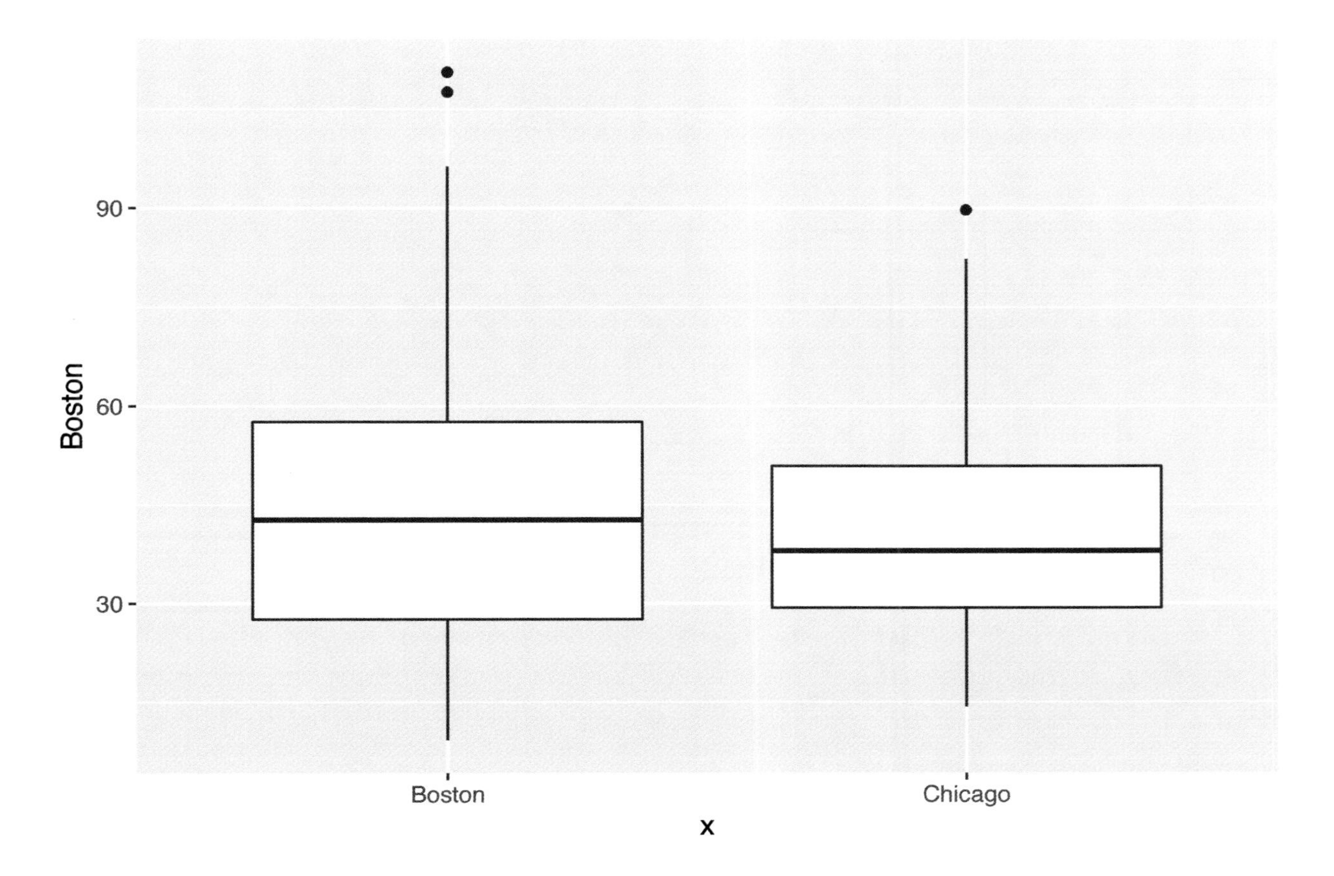

Figure 6.7: Box plot with an additional layer.

points for each winter, *and* there's important information stored in the *column names.* What do I mean by "information stored in the column names"? The only way you know which observation belongs to what city is by the column header, and that's definitely not tidy.

Here's what a tidy version of that data could look like, using a new tidy CSV file, snowdata_tidy.csv, in the data subdirectory:

```
##      Winter    City TotalSnow
## 1 1940-1941  Boston      47.8
## 2 1940-1941 Chicago      52.5
## 3 1940-1941     NYC      39.0
## 4 1941-1942  Boston      23.9
## 5 1941-1942 Chicago      29.8
## 6 1941-1942     NYC      11.3
```

That's not the most human-readable of formats, but it's very much a ggplot2-friendly structure. I'll show you how to reshape your data into a tidy data frame (sometimes called going from "wide" – with lots of columns – to "narrow" – with fewer columns – format) in Chapter 14.

Here's one advantage of tidy data: Instead of having to add each city manually in its own ggplot layer, you can simply use the City column for the x-axis, and ggplot will automatically make a separate plot for each, creating a plot as in Figure 6.9. If you add another city or change the format of a city's name, the exact same code will work.

```
ggplot(snowdata_tidy, aes(x = City, y = TotalSnow)) +
  geom_boxplot()
```

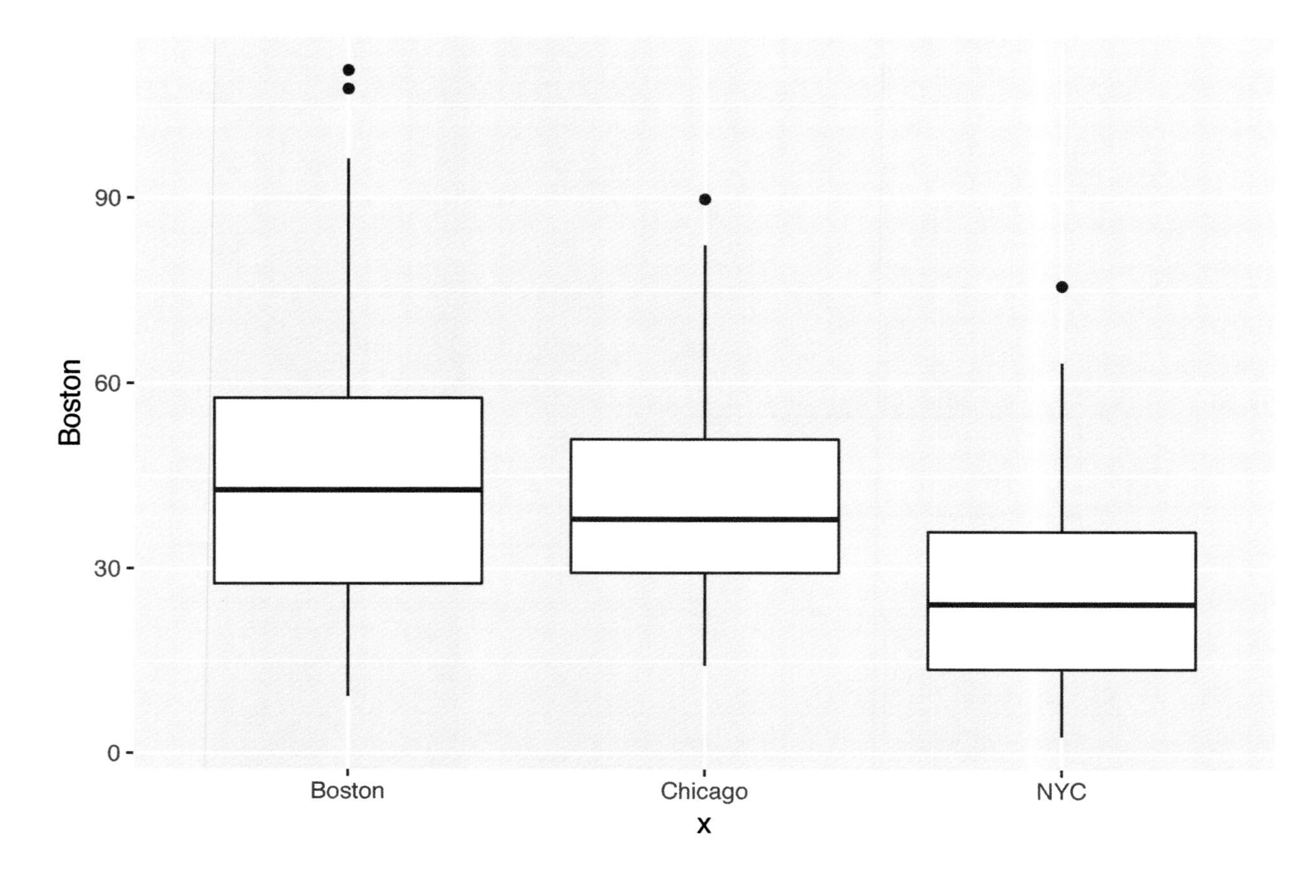

Figure 6.8: Box plot comparing three cities.

That works for a box plot. For a graphic that needs Winter as the x axis, such as a line chart or scatter plot, using City as the x axis won't work very well. How can you use the tidy version of this data to easily add one line per city? By adding a **group** aesthetic.

```
ggplot(snowdata_tidy, aes(x = Winter, y = TotalSnow, group = City)) +
  geom_line()
```

Try to ignore that the x-axis labels for Winter in Figure 6.10 are completely illegible. Instead, look at the plot. There are now 3 lines, one for each city ... but we can't see which one is which. We can fix that by coloring the lines by city, with the *`color` aesthetic*:

```
ggplot(snowdata_tidy, aes(x = Winter, y = TotalSnow, group = City, color = City)) +
  geom_line()
```

Still not the most useful of graphics, whether or not you can see the colors in the book's print edition, but we're getting closer.

However, please step back and make sure you understand this code before moving on to additional customizations. That first line is setting up the foundation for this visualization: Data comes from the snowdata_tidy data frame, x axis will be the Winter column, y axis will be the TotalSnow column, there will be one visualization series per city (that's the group = City), and each city will have its own color. But that first line doesn't say what kind of visualization this should be.

The second line of code says that we want a line graph.

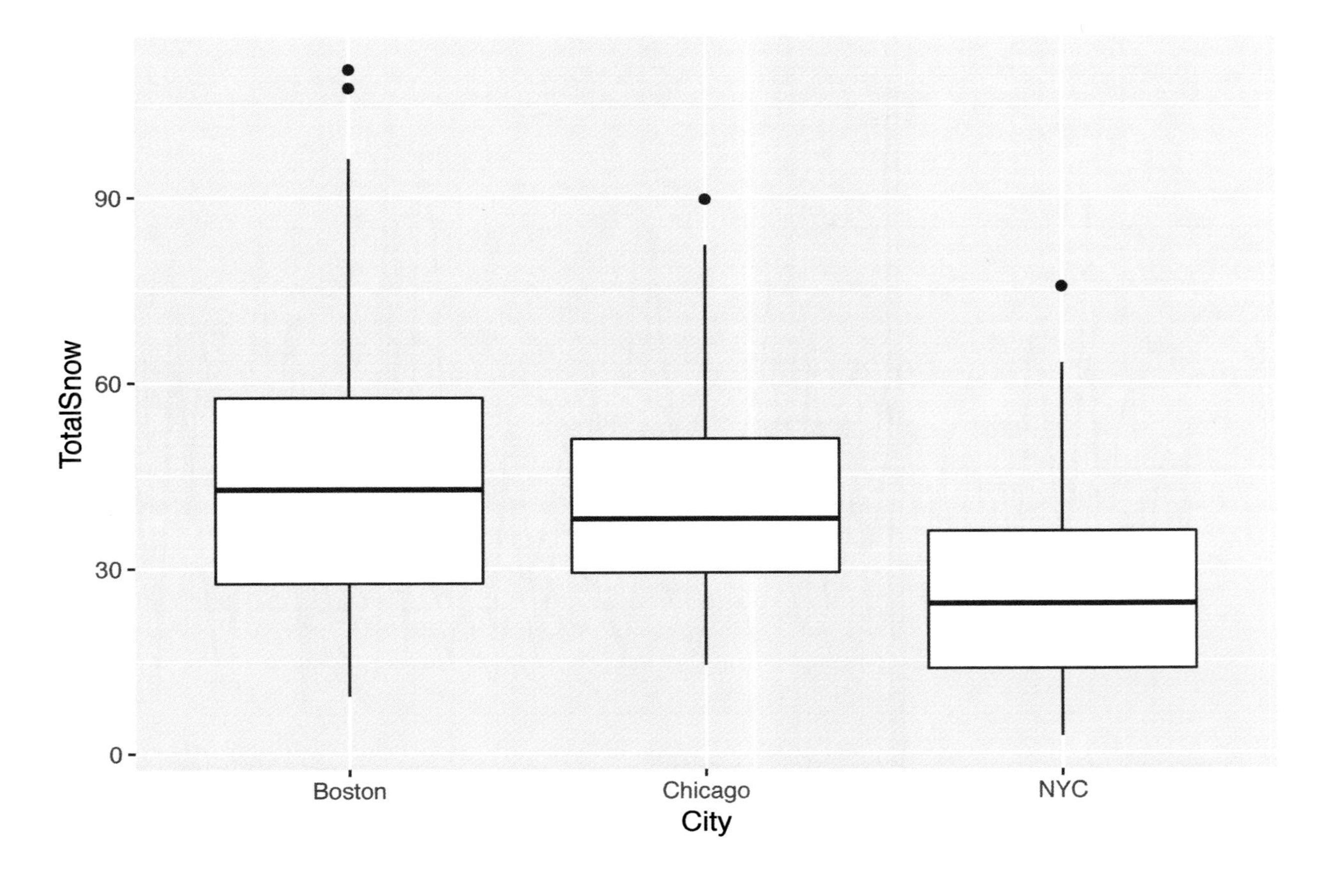

Figure 6.9: Box plot created with ggplot() and "tidy" data.

Right now the lines are a little tough to see. While I could make the line wider, for this data set the points with actual values are most interesting. I can add points to this line graph with another geom layer:

```
ggplot(snowdata_tidy, aes(x = Winter, y = TotalSnow, group = City, color = City)) +
  geom_line() +
  geom_point()
```

Do you see how you can build a graphic, layer by layer? (See Figure 6.12.)

Unfortunately, this particular graphic is kind of busy. So let's try using fewer data points, and just look at 21st-century winters by filtering the winters and then re-doing the graph:

```
snowdata_tidy21 <- filter(snowdata_tidy, Winter >= "1999-2000")
ggplot(snowdata_tidy21, aes(x = Winter, y = TotalSnow, group = City, color = City)) +
geom_line() +
  geom_point()
```

The x-axis labels in Figure 6.13 are still unreadable; I *will* get to rotating that text soon. Even more important, though, **in the next section I'll show you how you can keep all this complex ggplot2 syntax at your fingertips, so you don't have to memorize it or keep looking it up.**

Meanwhile, I'm guessing you're somewhat underwhelmed by the line graph as a way to understand this data. It's a good lesson on ggplot2 layers, but it's actually not a great way to look at trends in this specific data. Maybe a grouped bar chart by year would be more useful?

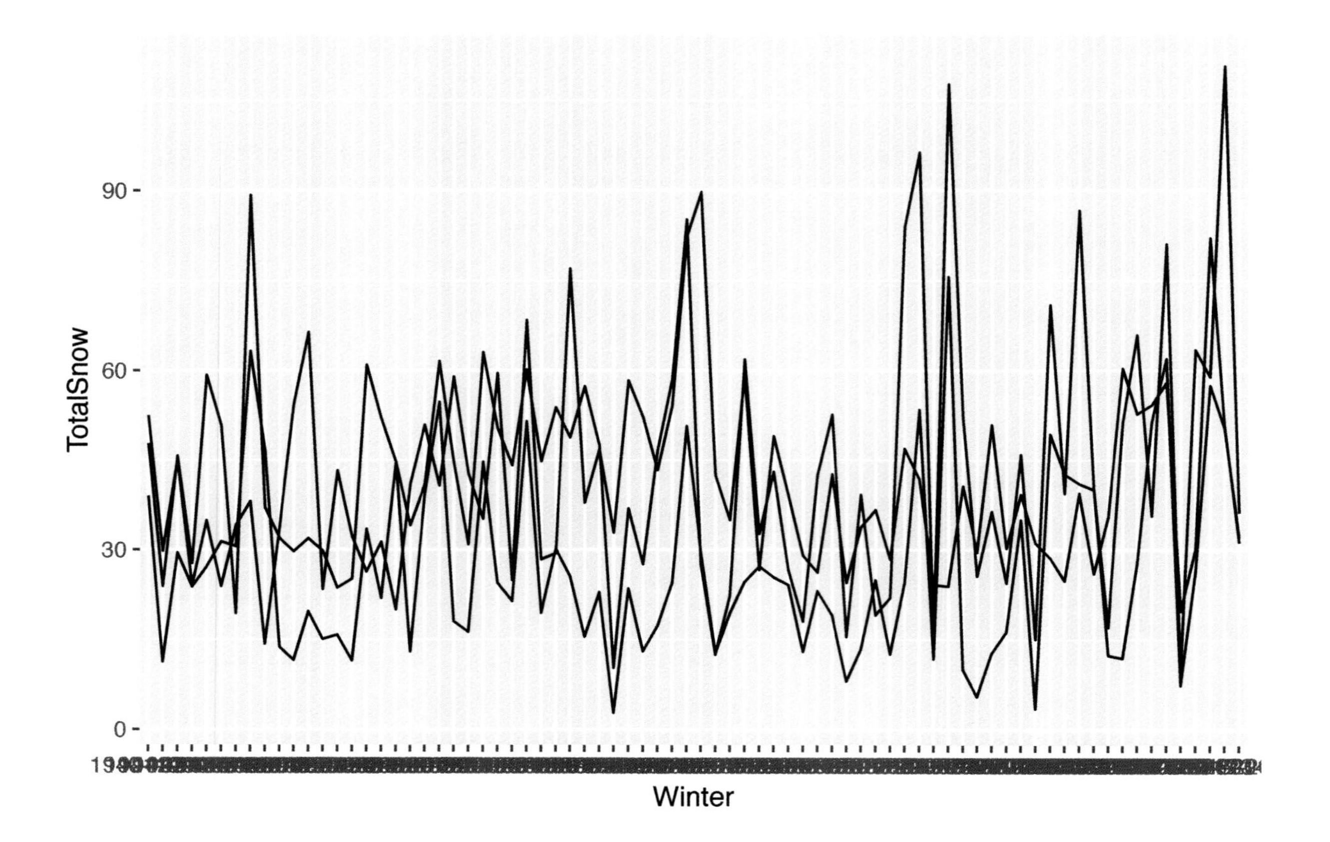

Figure 6.10: Line chart created with ggplot(). Ignore the x-axis text for now.

```
ggplot(snowdata_tidy21, aes(x = Winter, y = TotalSnow, group = City, color = City)) +
  geom_bar()
```

is close, but we get the same error as we saw with qplot earlier. With full ggplot2, though (as opposed to qplot()), we can fix this. There are two additions we need. In older versions of ggplot2, you had to add the somewhat unintuitive `stat = "identity"` within geom_bar() to tell ggplot2 that you don't want it counting number of items. Newer versions have a geom_col() function (for column chart), that has the default behavior you might expect.

In addition, we'll need to add `position = "dodge"` to either geom_bar() or geom_col() to say we want a *grouped* bar chart instead of a *stacked* bar chart.

One more issue: `color` is for lines. In a bar chart, `color` sets an *outline* color for the bars (makes sense if you think of it as the outline being a type of line). To *fill* the bars by color, you use `fill = City` and not `color = City`.

You may have to trust that this makes sense if you are operating under the grammar of graphics theory. I suspect it may be a little overwhelming, though, if you're new to programming. Hold on just one section longer, though, and you'll see how you can save all this information once within RStudio and then pull it up whenever you want.

Here's the basic grouped bar chart code, which produces the graph in Figure 6.14:

```
ggplot(data = snowdata_tidy21, aes(x = Winter, y = TotalSnow,
group = City, fill = City)) + geom_col(position = "dodge")
```

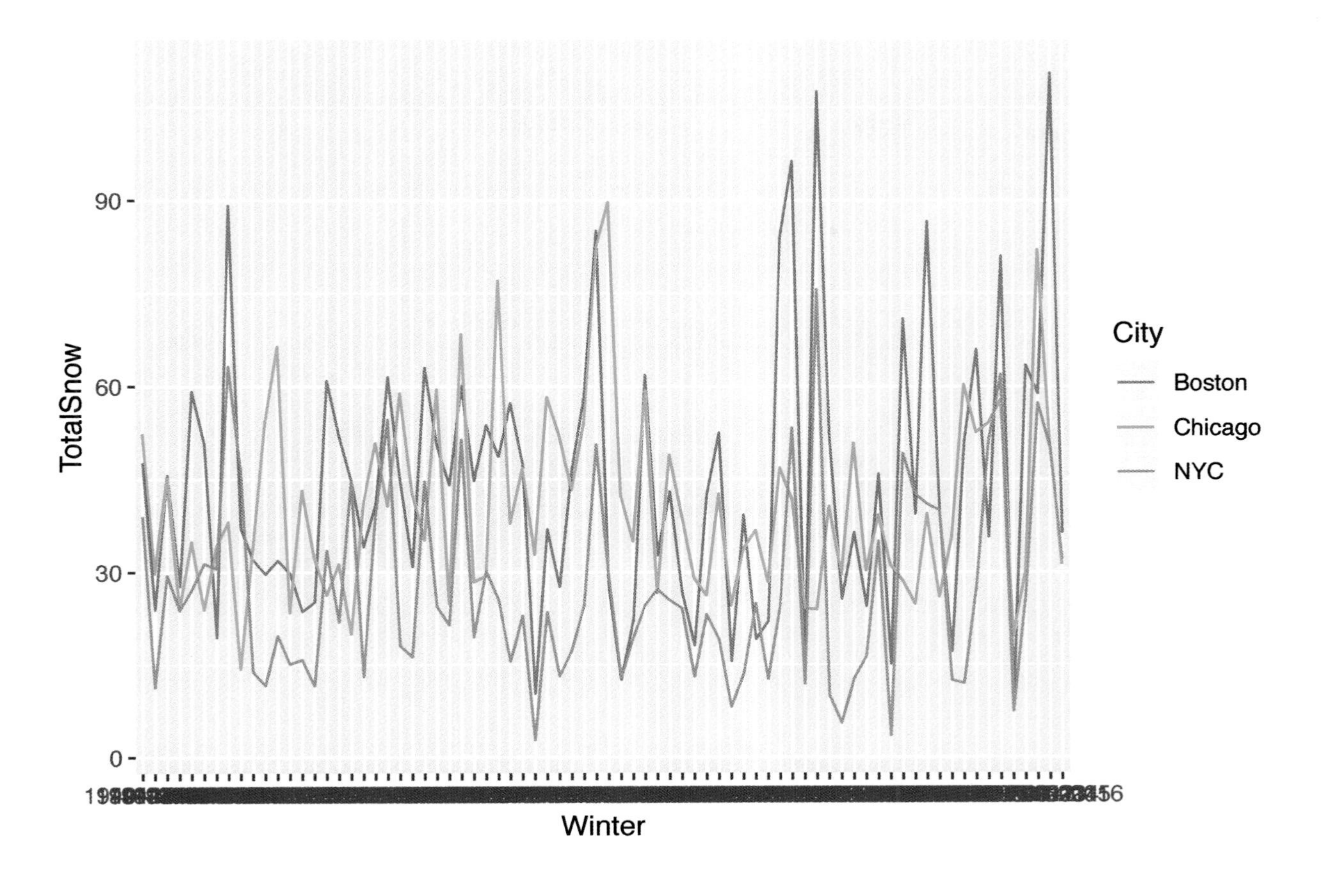

Figure 6.11: Adding a color aesthetic.

Or just

```
ggplot(snowdata_tidy21, aes(Winter, TotalSnow, group = City, fill = City)) +
  geom_col(position = "dodge")
```

6.9 Code snippets to the rescue

There's a *lot* of syntax to remember with ggplot2, and not all of it is intuitive. That's often the case when learning a programming language. And even after you're proficient and *do* remember much of the syntax, some code phrases can get tedious to type. That's why most IDEs, as part of their "make life easier for programmers" design, include some sort of macro-like functionality so users can quickly look up code chunks, insert one, and then "fill in the blanks."

This will be easier to see with an example. In base R, there's a function called lapply that applies a function over each element in a list (another data type that we'll cover later list). RStudio has a built-in code snippet for lapply. Type `lapp` into either your source code window at top left or interactive console at bottom left, and you should see an autocomplete choice for the lapply *snippet* as well as the lapply *function* in base R. Hit the tab key to select the snippet, and you should see something like this:

```
lapply(list, function)
```

The first argument for lapply, list, appears already defined. It's a placeholder variable, which you can replace by typing the name of your own list. Then hit tab, and the cursor will jump to define function and you can

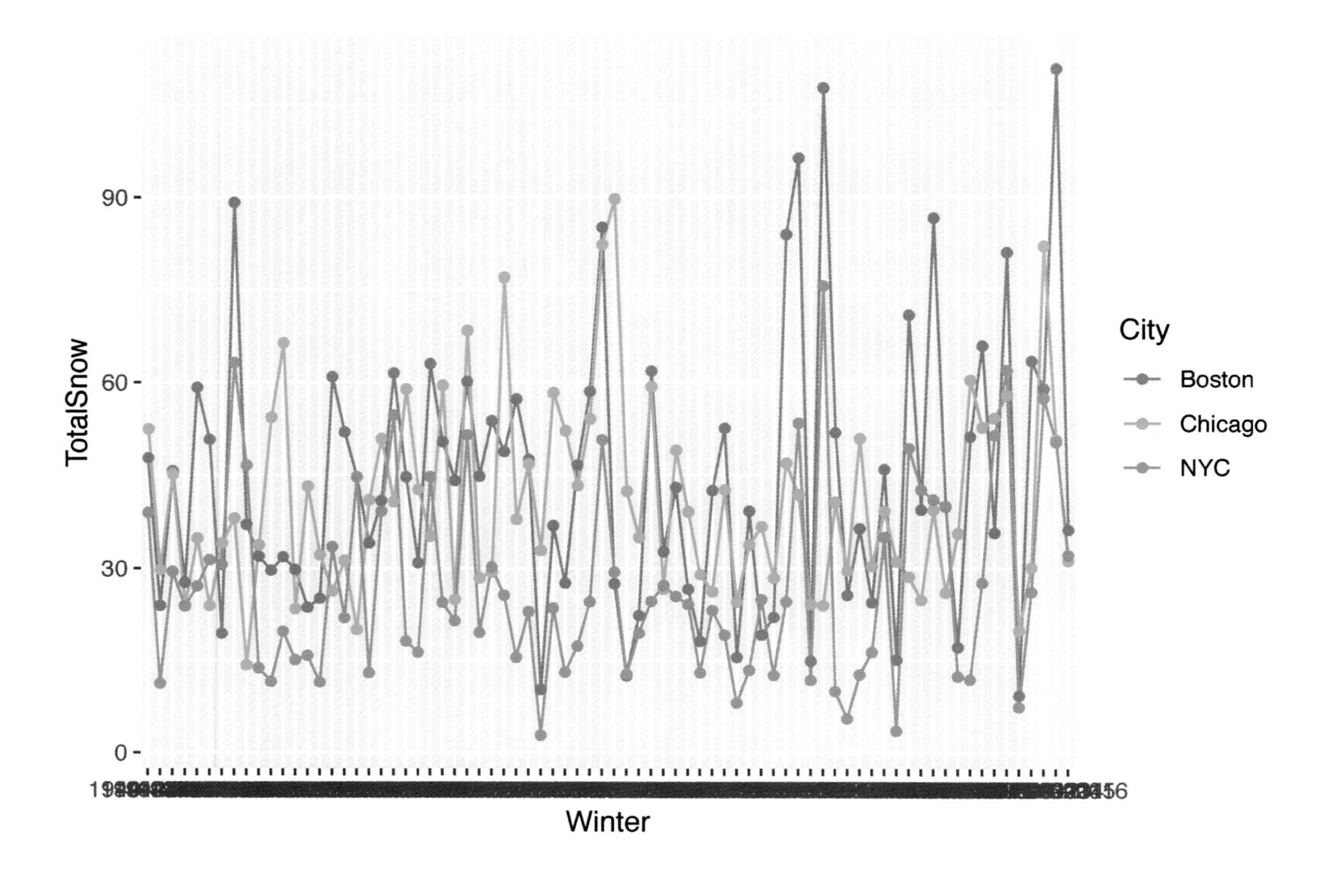

Figure 6.12: Line graph with points using ggplot().

start typing to replace that one. This is very handy for both remembering the exact format of a function and having the fill-in-the-blank framework.

But along with built-in snippets, you can create and save your own, by adding them to the snippet file within RStudio. The RStudio graphical-user-interface way to access that file is with the menu commands Tools > Global Options > Code (see Figure 6.15). Then click the Edit Snippets near the bottom of the dialog box to bring up the file.

Clicking through that chain of menu commands can get tedious. Fortunately, there's also a function in the usethis package that immediately opens that snippet file for editing: `edit_rstudio_snippets()`.

Like an R script, this is just a plain text file. However, snippets have a fairly strict structure. If you look at the lapply snippet, it's defined like this:

```
snippet lapply
	lapply(${1:list}, ${2:function})
```

A more generic snippet syntax looks something like this:

```
snippet snippetname
	myfunction(${1:variable1name}, ${2:variable2name})
```

The first line starts with the word **snippet**, a space, and then the name you choose for your snippet.

The code for the snippet goes immediately under its name, and *each line of the snippet code MUST be indented with a tab* (not spaces). Every fill-in-the-blank variable is numbered. The first fill-in-the-blank variable would take the format `${1:variable1name}` – that's a dollar sign, an opening curly brace, the

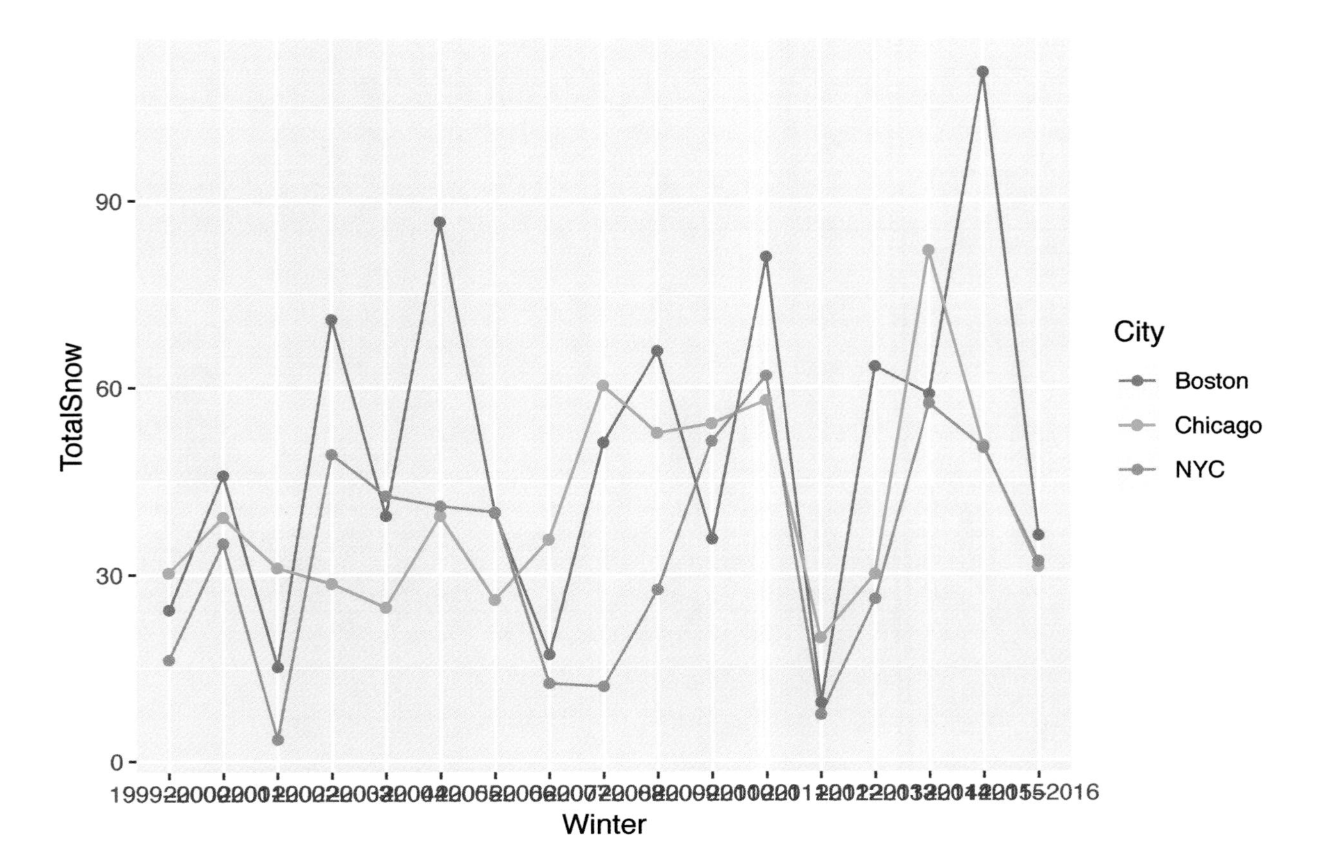

Figure 6.13: Graphing 21st-century snowfall by winter in three cities.

variable number (in this case 1), a colon, and then whatever name you choose for the variable. The more descriptive the better – as you see in the lapply snippet, the first variable is named list and the second is named function, so users know right away how to fill them in.

The grouped bar chart code was

```
ggplot(snowdata_tidy21, aes(Winter, TotalSnow, group = City, fill = City)) +
  geom_col( position = "dodge")
```

If you want to save that as a snippet to use for future bar charts, first you need to give it a name – preferably one that you'll remember or that's easy to find within RStudio. I start all my custom ggplot2 snippets with `myg_` so that whenever I want to use one, I can start typing myg_ and see a dropdown list of available choices. In my naming convention, I'd call this `myg_barplot_grouped` so that if I'm working specficially on a bar chart, I can type `myg_barplot` and see all my barplot snippet options.

Next, look at that code and see which parts should be turned into variables that you'd replace if you were making the same chart with different data. I see four: snowdata_tidy21, Winter, TotalSnow, and City (which appears twice).

Here's a code-snippet version of the grouped bar chart code:

```
snippet myg_barplot_grouped
    ggplot(${1:mydataframe}, aes(${2:xcolname}, ${3:ycolname}, group = ${4:groupbycolname},
      fill = ${4:groupbycolname})) +
      geom_col(position = "dodge")
```

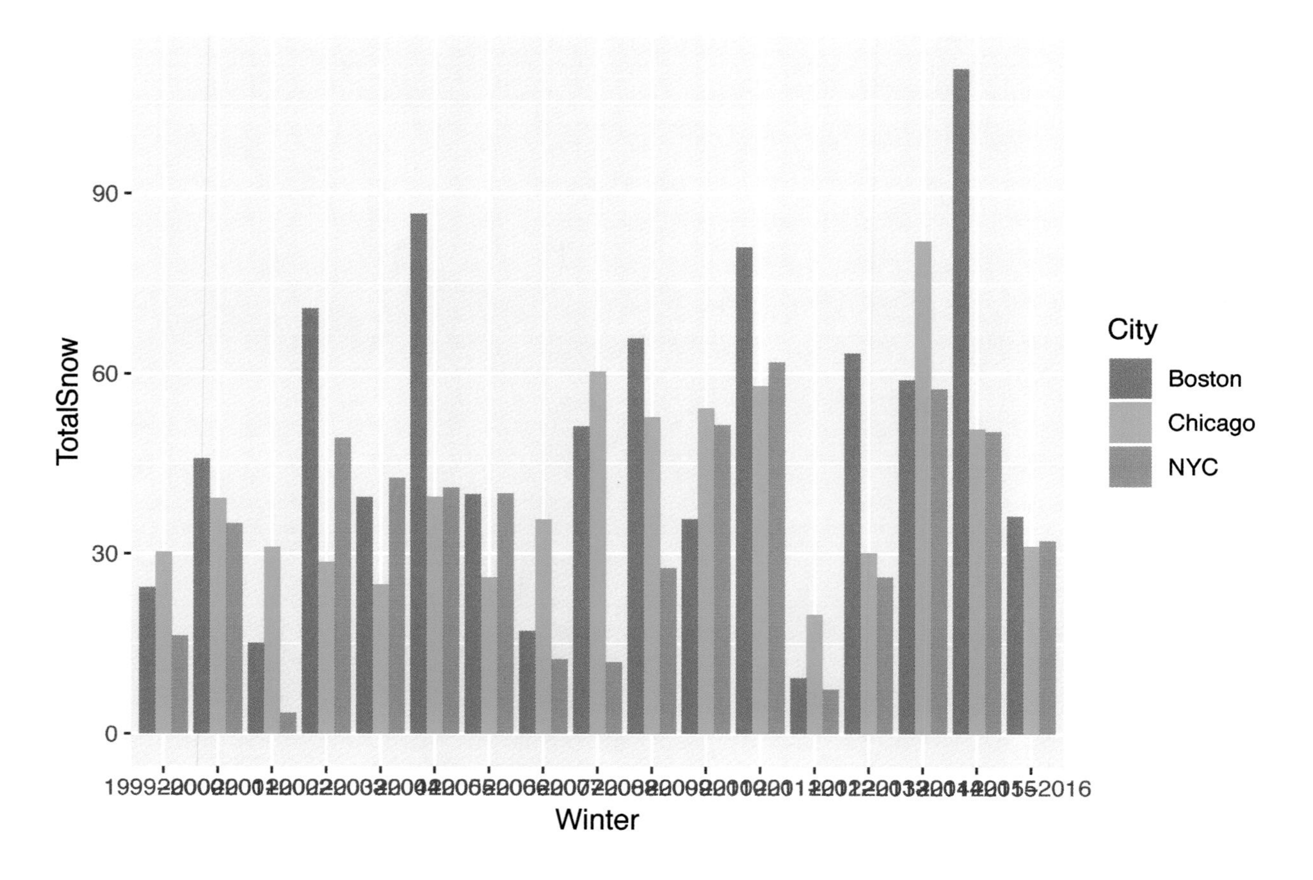

Figure 6.14: Grouped bar chart with ggplot().

By repeating `${4:groupbycolname}` twice in the snippet, you'll only have to type it once when coding: What you've typed will appear in both places. Call the variables by names that will make sense to you, and remember to indent each line with a tab to start. You can add space *after* the tab, but the lines must start with a tab.

Now save the file, hit OK, go back to your R script window or console, and start typing myg_ (or the first few letters of whatever you called your snippet). Select the new snippet with the tab key, and you should see something like Figure 6.16, with the first variable defined. Type in the name of a data frame and hit the tab key, and the cursor will jump to the next variable that needs to be defined. Keep doing that until you're finished with variables, and you should see the 4th variable name appear twice when you type it in.

Won't that be easy the next time you want to make a basic grouped bar chart?

6.10 Presentation-quality graphics

Code snippets become even more useful after you start building some presentation-quality graphics with customized colors, backgrounds, axes, titles, and so on. A nice-looking ggplot2 graphic can have 10 or more lines of code. Once you get a graphic to look exactly the way you want it, just remember to save the code as a snippet (with a name you can remember or at least find again) for re-use.

What type of visualization might you want for this weather data to run with a story, broadcast, or report? One common way of putting an extreme weather event in some context is to show it in a top 10 bar chart. Here's how you might put the 2014-15 Boston Snowpocalypse into a 10-snowiest-winters graphic.

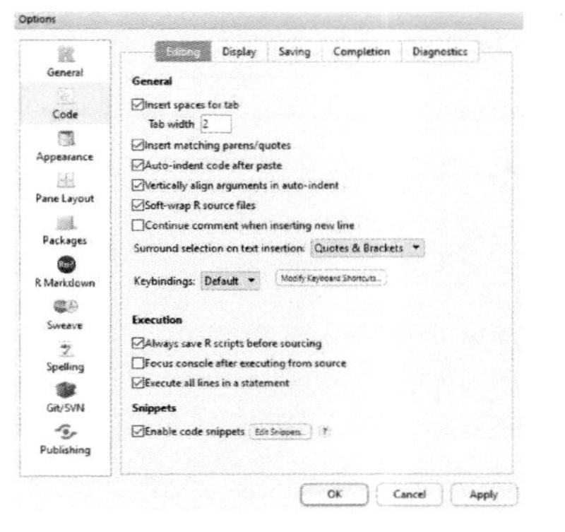

Figure 6.15: Editing a snippet file within RStudio

```
ggplot(mydataframe, aes(xcol, ycol, group = groupbycol, fill = groupbycol)) +
  geom_bar(stat = "identity", position = "dodge")
```

Figure 6.16: Code snippet with the first variable defined

Before starting to write this or any fairly complex R code, try to think of the *steps* you'll need your code to perform. Programmers call this pseudo code. The idea is to sketch out what steps need to be done to accomplish a task *before actually starting to write code.*

For this task, you might want to:

Create a sorted data frame of the top 10 snowfalls:

1. Get the Boston snowfall data
2. Sort it in descending order
3. Filter just the top 10

Then, create a bar chart of that data with ggplot2:

1. Create the foundation ggplot layer
2. Add the geom_bar() layer
3. Customize bar colors
4. Simplify background
5. Add title and data source
6. Make any additional tweaks as needed, such as customizing axes (and finally rotating the x-axis labels)

For the first task, creating a data frame of top 10 winters, one idea might be to arrange the data in descending order of snowfall, and then store the first 10 rows. That will work, but only if you're not worried about the 10th row being tied with row number 11. If there is a tie at the 10th spot, only one row will be included.

dplyr has a function to deal with ties in rankings: top_n(). Its format is `top_n(mydataframe, number of rows to return, column you want to rank)`. So, to get the top 10 rows by TotalSnow from the bostonsnow data frame, the syntax is `top_n(bostonsnow, 10, TotalSnow)`. Here's the full "piped" dplyr code to create the new data frame:

```
boston10 <- bostonsnow %>%
    top_n(10, TotalSnow) %>%
    arrange(desc(TotalSnow))
```

The first line copies the bostonsnow data frame into a variable called boston10, and pipes that result to the second line.

Line 2 takes that boston10 data frame and applies the top_n function to it, extracting the 10 rows with the highest TotalSnow values (more than 10 rows if one or more rows after row 10 are tied with #10). That new data frame result is piped to the third line, which orders it from highest to lowest by TotalSnow.

Now it's time to learn how to do some ready-to-share visualizations with full ggplot2.

We already went over the first two tasks for building a graph: the foundation layer `ggplot(mydf, aes(x = xcolname, y = ycolname)` and the geom layer, in this case geom_bar(). This second line will use the syntax `geom_bar(stat = "identity", fill = "mycolorchoice")`.

Reminder: That first geom_bar() argument, `stat = "identity"`, says that the bar chart *shouldn't* use the default of counting items for the y axis, but instead *should* use *values* from mydf$ycolname for the y axis. The fill argument is pretty self-explanatory, setting the bar colors. Another optional argument, `color`, would set an *outline color* for the bars. (As a beginner, it's easy to assume that `color = "blue"` will make the bars blue, but it only *outlines* them in blue).

ggplot2 defaults to a rather distinctive grey background for its plots. If you'd prefer something else, you can either build the background manually (something I wouldn't recommend while still learning the basics) or use a pre-built alternative theme. Type `theme_` in the RStudio console and you'll see several different options. If those aren't enough, you can install additional themes with packages like ggthemes. Try some of the different themes and see if there's one you like.

I'll use the minimal theme, which uses a white background and a much lighter grid, by adding `+ theme_minimal()` to the graph. For an even *more* spartan theme, install and load the ggthemes package. There are themes that mimic the styles of some well-known publications like the Wall Street Journal, and one called theme_tufte(), based on graphics theories of dataviz pioneer Edward Tufte. Tufte advocates removing distractions when displaying information such as garish colors and, yes, background grids.

hrbrthemes is another package with ggplot2 themes, primarily aimed at typography. It's quite well-regarded; if you'd like to take it for a spin, install it with `githubinstall::gh_install_packages("hrbrthemes")` (if you are asked to choose among multiple versions, select hrbrmstr/hrbrthemes) and then try themes such as theme_ipsum().

So far, using theme_minimal(), the graph's first three lines look like this:

```
ggplot(data = boston10, aes(x = Winter, y = TotalSnow)) +
  geom_col(fill = "dodgerblue4") +
  theme_minimal()
```

Aside: If you're wondering how I knew I could use a color called "dodgerblue4" for Figure 6.17, R's `color()` function prints out all 657 available colors by name and `demo(colors)` displays them. However, there are some nice PDFs online that you might find more helpful, such as Columbia University assistant professor Tian Zheng's R Colors at www.stat.columbia.edu/~tzheng/files/Rcolor.pdf. And if you don't want to type that URL and have already downloaded this book's GitHub repository files, open the `booklinks.html` file in your browser and scroll down to the section on Chapter 6.

You can also use color Hex codes instead of R's built-in named colors when specifying colors for your charts and graphs, such as `geom_bar(stat = "identity", fill = "#0072B2")`.

To make it easier to try different options in a ggplot2 graphic, you can store the first few lines in a variable and then add new options to that base, such as:

```
p <- ggplot(data = boston10, aes(x = Winter, y = TotalSnow)) +
  geom_col(fill = "dodgerblue4")

p + theme_minimal()
p + theme_tufte()
p + theme_economist_white()
```

and so on.

As of ggplot2 version 2.0, you can add a title, subtitle, and caption to a ggplot graph with the labs() function:

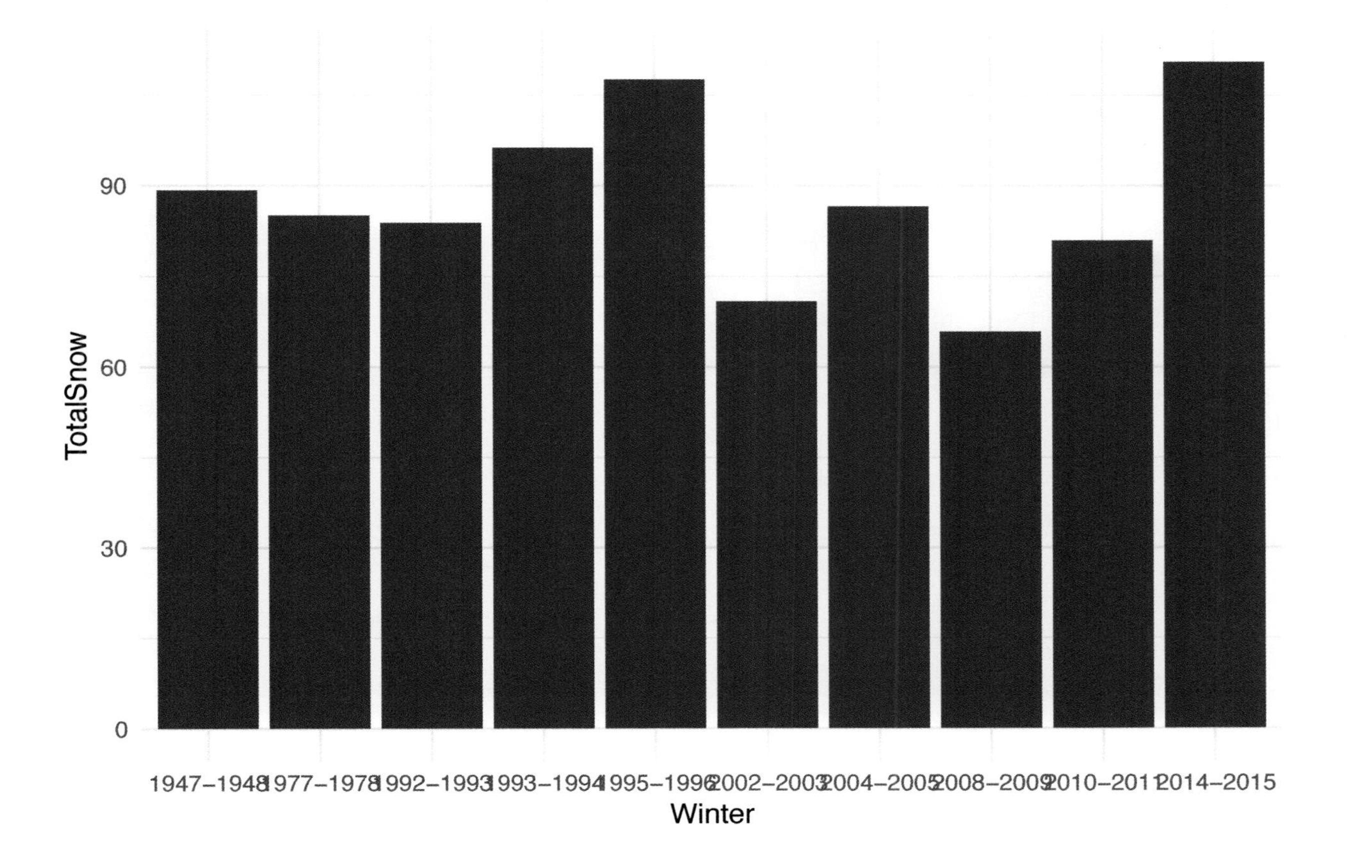

Figure 6.17: ggplot2 bar chart using the minimal theme.

```
ggplot(data = boston10, aes(x = Winter, y = TotalSnow)) +
  geom_col(fill = "dodgerblue4") +
  theme_minimal() +
  labs(title = "Top 10 Snowiest Winters in Boston",
  subtitle = "Total snowfall in inches",
  caption = "Source: National Weather Service")
```

The graph in Figure 6.18 still needs some fixes, such as making the x axis more legible. But before I turn to that, there's an important change I need to make for the story I'd like to tell: I want the bars ordered *from highest to lowest* (just like I arranged my data frame), so it's easy to spot the highest, next-highest, 3rd-highest, etc. totals. Unfortunately, ggplot defaults to alphabetical order when the x axis is character strings.

I struggled with this problem when first learning ggplot(), but there are actually a couple of simple answers. One is to reorder the x-axis graph itself; the other is to turn the x-axis vector of character strings into an ordered factor.

For this particular graph, reordering within the ggplot2 code makes sense. You can do this using the syntax `aes(x=reorder(myxcolname, -myycolname))`. This code says "Use myxcolname as the data for the x axis, but reorder it *based on the values of the myycolname column when arranged in descending order.*

```
ggplot(data = boston10, aes(x = reorder(Winter, -TotalSnow), y = TotalSnow)) +
  geom_col(fill = "dodgerblue4") ...
```

Or, you can use Hadley Wickham's forcats package, which was designed for dealing with categorical variables.

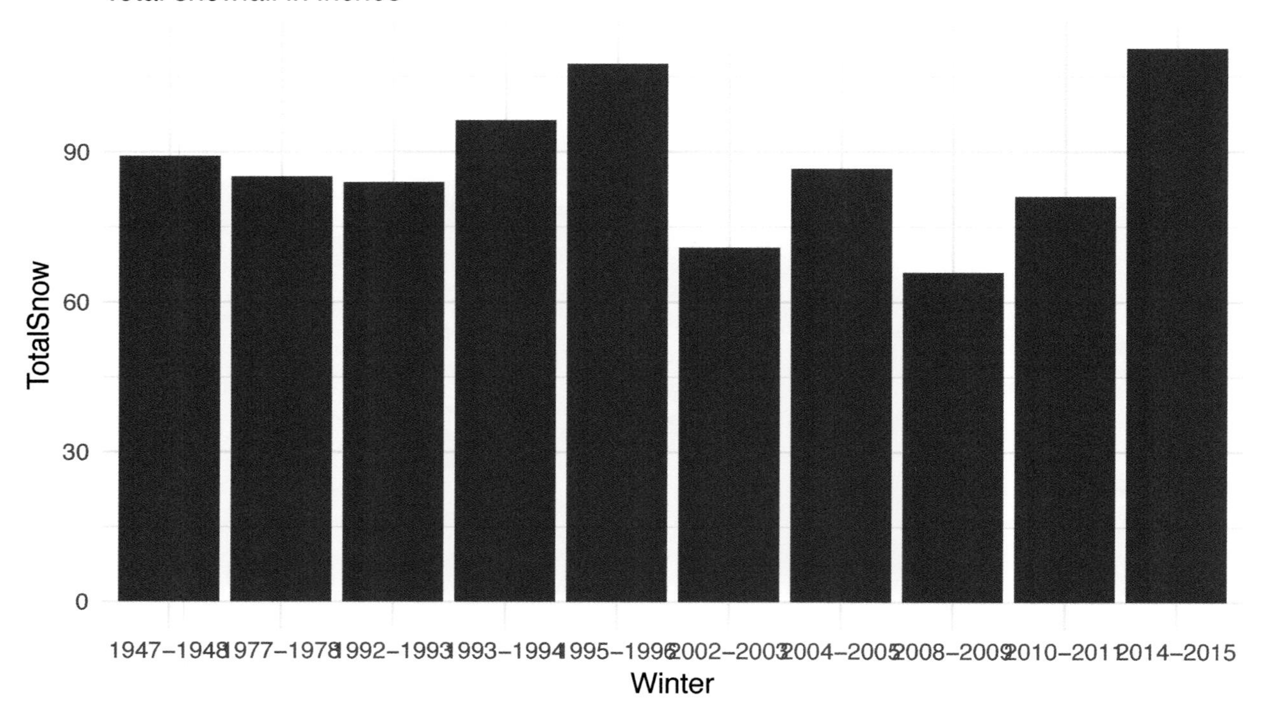

Figure 6.18: Adding title and subtitle.

forcats' fct_reorder() function was made for easily reordering one variable by another, using the syntax `fct_reorder(vectorToReorder, vectorToReorderBy)`. Here's how you'd use it in the plot (minus the other refinements):

```
ggplot(boston10, aes(x=fct_reorder(Winter, TotalSnow), y=TotalSnow)) +
  geom_col()
```

I don't see a huge advantage of fct_reorder() compared with base R's reorder, but forcats has other functions making it easy to deal with factors. And, you're likely to run into it when looking at code by people throughout the tidyverse.

Sometimes, though, factors would be the better option than reordering within ggplot2. For example, if you were graphing data by month and wanted to show trends over time, you probably wouldn't ever want your x axis to be alphabetically ordered as April, August, December etc. or rearranged based on the y-axis data.

Creating a factor from a character vector isn't the most intuitive of R operations, but it does make some sense once you see how it's done. I'll demonstrate with R's built-in vector of month names, `month.name`, which are character strings. Here's how you'd turn those into ordered factors in base R:

```
month_factor <- factor(month.name, levels = month.name, ordered = TRUE)
```

The format is `factor(myvector, levels = orderedLevels, ordered = TRUE)`. By ordered levels, I mean a vector showing each item in the proper order. In this case, month.name already does that.

Now take a look at the result:

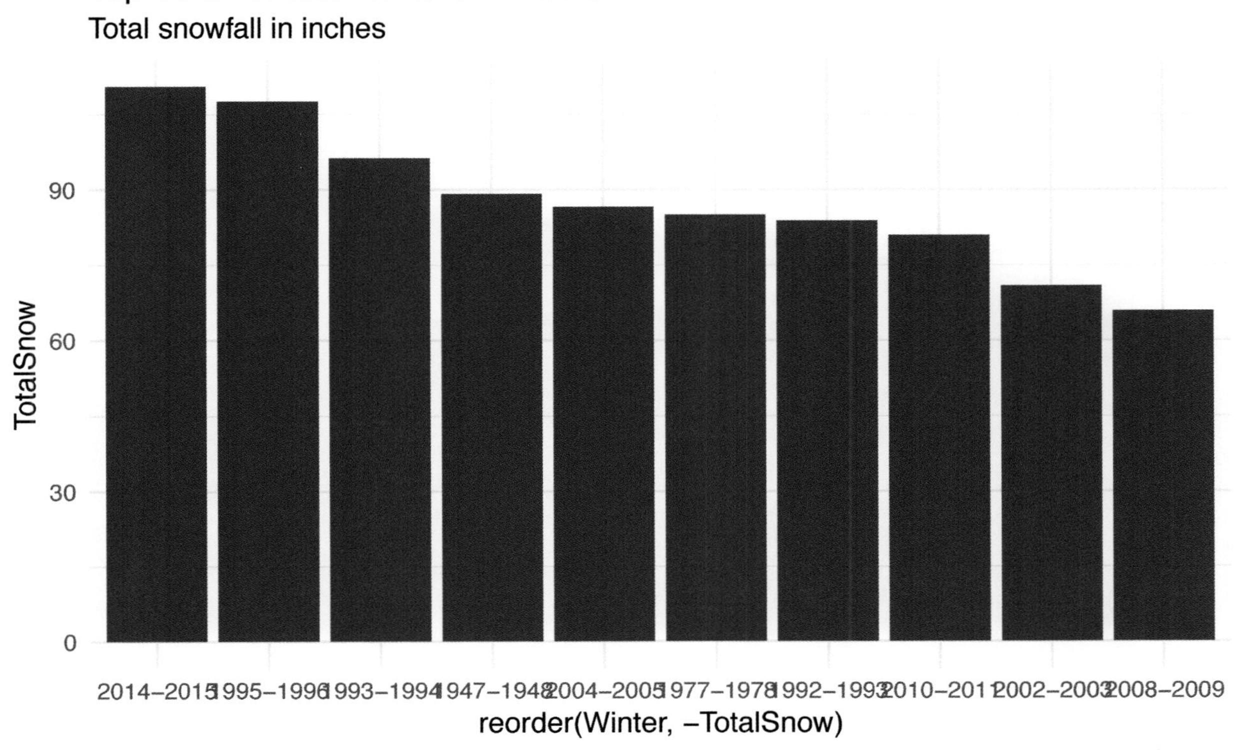

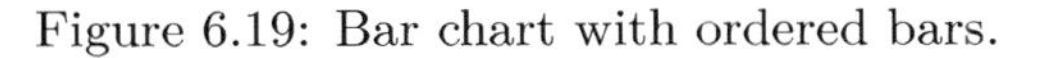
Figure 6.19: Bar chart with ordered bars.

```
month_factor
```

```
##  [1] January   February  March     April     May       June      July
##  [8] August    September October   November  December
## 12 Levels: January < February < March < April < May < June < ... < December
```

The elements no longer have quotation marks around them, and the factors are clearly ordered based on what's specified in the levels argument (in this case the original order).

This code creates a factor from the boston10$Winter vector:

```
boston10$Season <- factor(boston10$Winter, levels = sort(boston10$Winter))
```

(Unless memory is an issue, it's always better to create a new column when making changes in a data frame, just in case something goes wrong.) You can see the graph is properly ordered now, since my levels are the sorted order of the Winter column.

In the code below for Figure 6.20, I'm storing the first part of the graph in the variable myplot and then displaying that variable so I can keep adding to the graph without re-typing all the code.

```
myplot <- ggplot(boston10, aes(x=reorder(Winter, -TotalSnow), y=TotalSnow)) +
  geom_col(fill = "dodgerblue4")  +
  theme_minimal() +
  labs(title = "Top 10 Snowiest Winters in Boston",
    subtitle = "Total snowfall in inches",
    caption = "Source: National Weather Service")
```

```
myplot
```

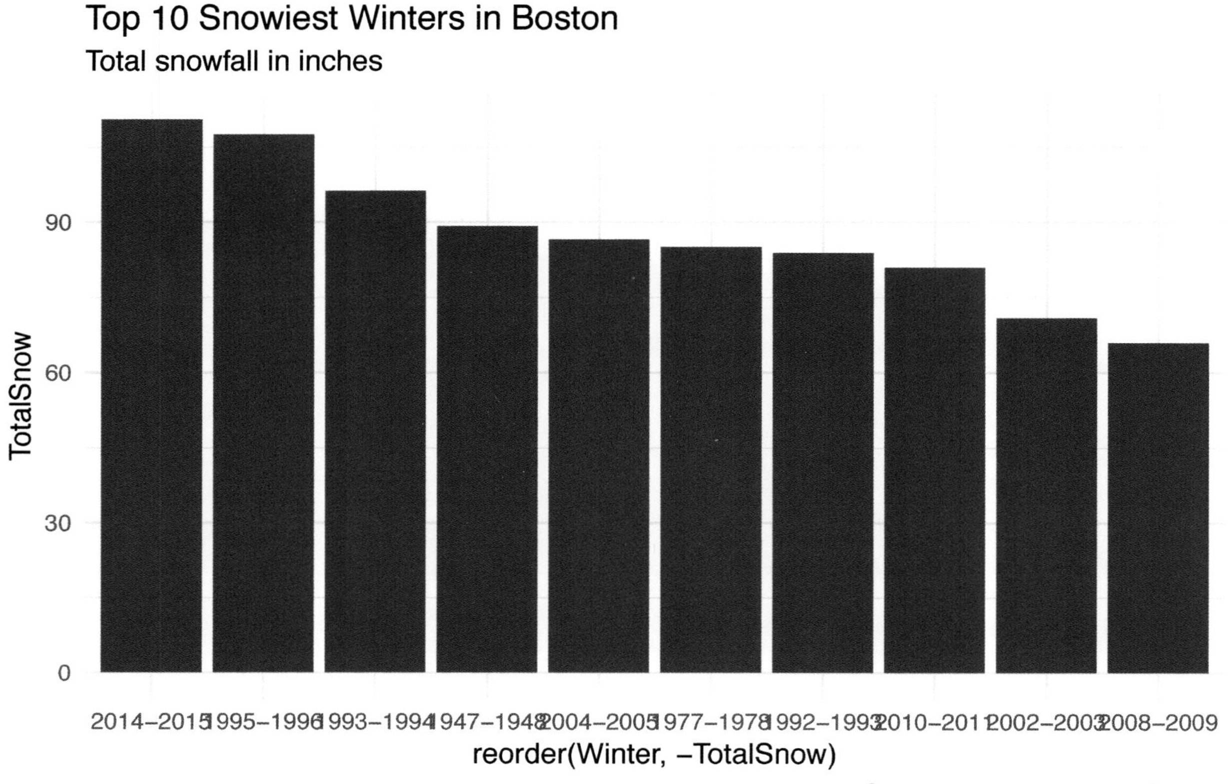

Figure 6.20: ggplot2 graph stored in a variable and then displayed.

6.10.1 Tweaking the axes

It's finally time to make that x-axis text more legible. I cheated on this one when learning ggplot2 and simply looked at Winston Chang's Cookbook for R online at http://www.cookbook-r.com/Graphs/. In the axes section, he shows rotating the x-axis text 90 degrees with 'theme(axis.text.x = element_text(angle = 90, vjust = 0.5))

I usually prefer a 45-degree rotation, as in Figure 6.21:

```
myplot + theme(axis.text.x = element_text(angle = 45, vjust = 0.5))
```

That's fine except the axis labels need to be shifted a bit to the left. That's an hjust amount, and after a fair amount of experimentation, I settled on

```
myplot + theme(axis.text.x = element_text(angle = 45, vjust = 1.2, hjust = 1.1))
```

I would never remember this code to use again in future graphs. That's why I added this as a code snippet, to create x-axis text as in Figure 6.22:

```
snippet myg_axis_rotate_text
    theme(axis.text.x = element_text(angle = 45, vjust = 1.2, hjust = 1.1))
```

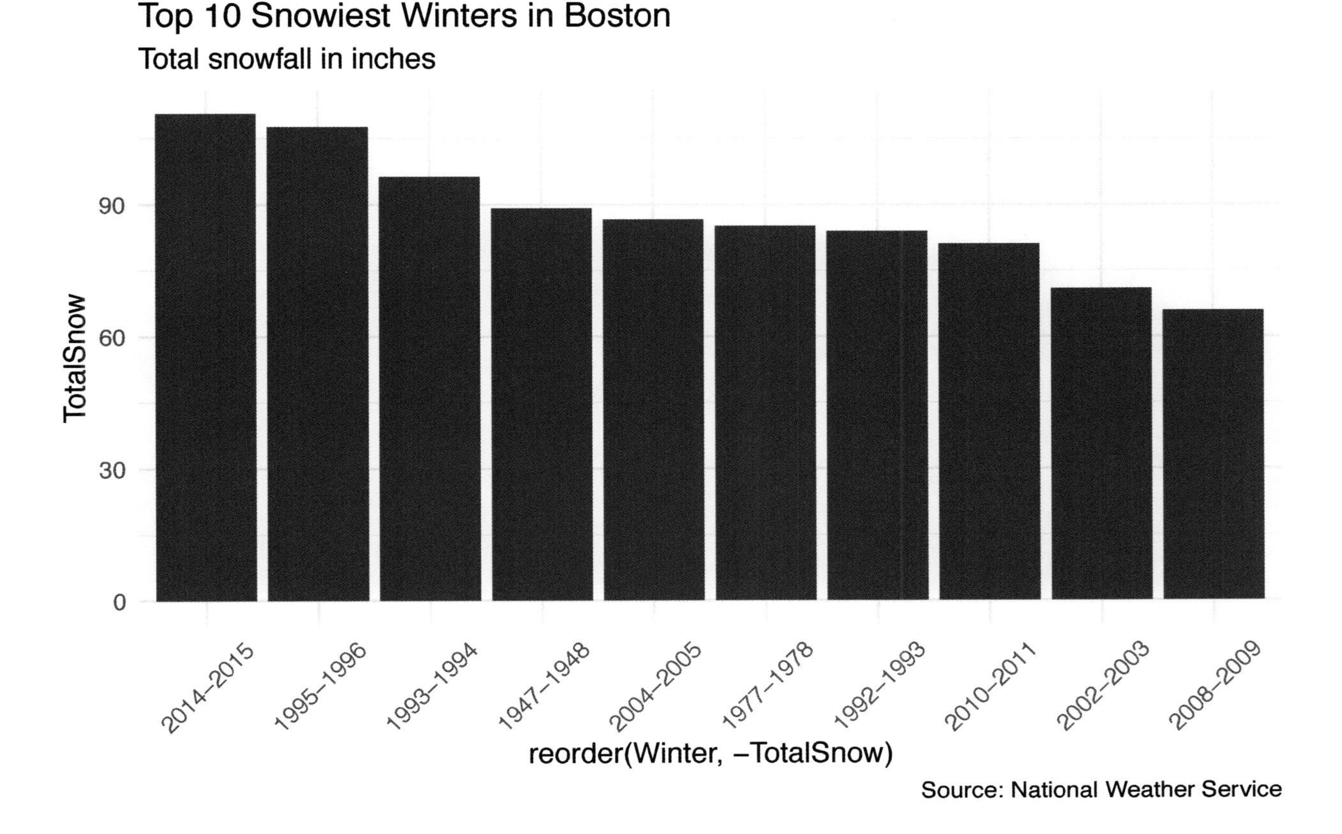

Figure 6.21: Graph with x-axis text rotated 45 degrees.

Even if I start typing `myg_rotate` instead of `myg_axis`, this snippet should come up in the dropdown list and I can add it to a graph (remembering to include a `+` at the end of the previous line of code, and not a `%>%`).

Last few tweaks: I'll change the axis titles and rotate the Winter labels. `xlab("My graph title")` sets the x-axis title, `ylab()` does the same for the y axis. reorder(Winter, -TotalSnow) is a terrible axis label, while I don't think I need a y-axis label at all:

```
myplot <- myplot +
  xlab("Winter") + ylab("")
```

You can also put the x and y text within the lab() function instead of using the separate xlab() and ylab() functions: `labs(title = "Top 10 Snowiest Winters in Boston", subtitle = "Total snowfall in inches", caption = "Source: National Weather Service", x = "Season", y = "")`. It's up to you whether you prefer to keep all your labels together or build your labeling line by line.

6.10.2 Make this bar graph a code snippet

Once you've got the graph as you like it, make a code snippet so you can easily re-use it. Look at the final graph code and see what text should be variables that will change when your data changes:

```
ggplot(data = boston10, aes(x = reorder(Winter, -TotalSnow), y = TotalSnow)) +
    geom_col(fill = "dodgerblue4") +
      theme_minimal() +
      labs(title = "Top 10 Snowiest Winters in Boston",
```

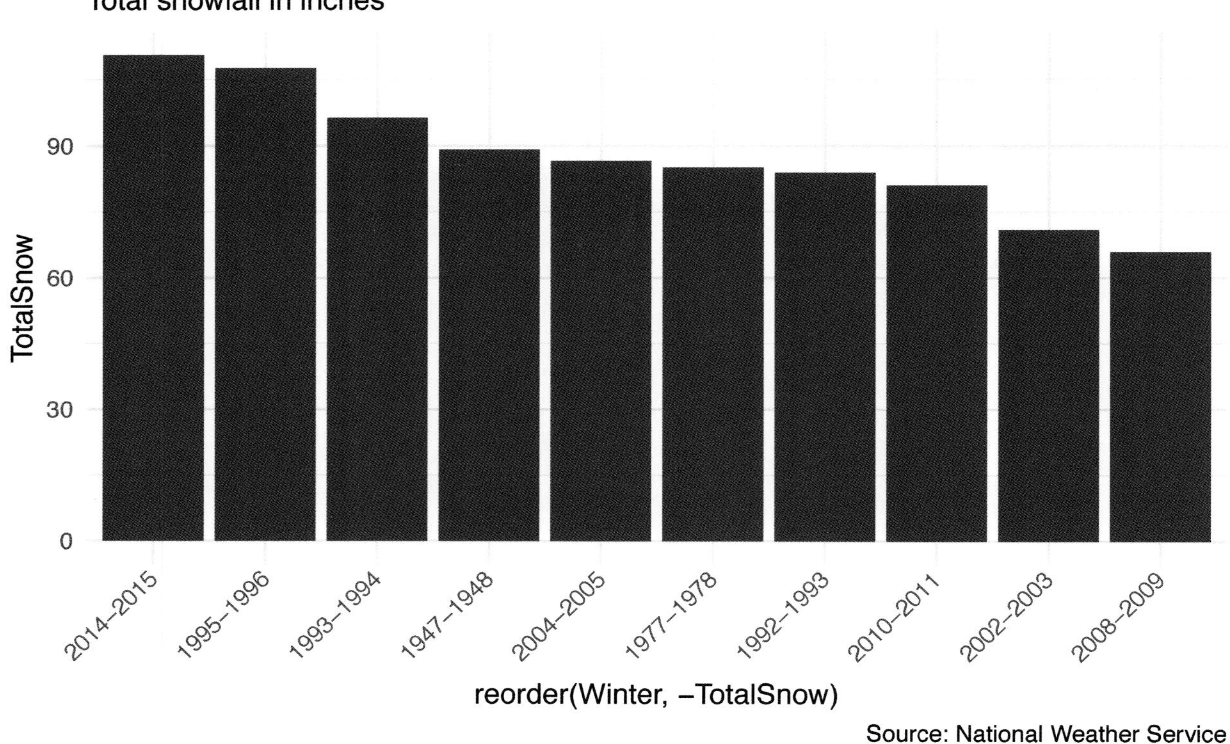

Figure 6.22: Graph with x-axis text rotated and properly positioned below the bars.

```
      subtitle = "Total snowfall in inches",
      caption = "Source: National Weather Service",
      x = "Winter", y = "") +
  theme(axis.text.x = element_text(angle = 45,
       vjust = 1.2, hjust = 1.1))
```

The changeable options are the data frame, x- and y-axis column names, x- and y-axis label names, title, subtitle, and caption. So, here's the graph in snippet form to add to an RStudio snippet file:

```
snippet myg_barplot_basic_reorderXaxis
    ggplot(data = ${1:mydf}, aes(x = reorder(${2:xcolname}, -${3:ycolname}),
    y = ${3:ycolname})) +
      geom_bar(stat = "identity", fill = "dodgerblue4")  +
      theme_minimal() +
      labs(title = "${4:mytitle}", subtitle = "${5:mysubtitle}",
      caption = "${6:mycaption}", x = "${7:myxaxislabel}",
      y = "${8:myyaxislabel}") +
      theme(axis.text.x = element_text(angle = 45, vjust = 1.2, hjust = 1.1))
```

This is a fairly complex snippet, so don't get discouraged if it takes you a few tries before you get the syntax exactly right. It took me a few tries to type it correctly, and I've been writing snippets for quite awhile. Pay careful attention to closing your curly brackets, and making sure the entire variable - including the dollar sign, open brace, number, colon, variable name, and closing brace - is inside quotation marks if you've got a character string that needs to be inside quote marks.

It's much easier to copy and paste someone else's snippet, but I strongly urge you to try to create a few snippets of your own. Snippets are one of the biggest time-saving options that RStudio (or any other IDE) offers. Investing a little time in learning how to use them will save you a ton of time in the future.

6.10.3 Two more tweaks.

There are many more ways to customize ggplot2. Here are a couple to start off:

- **Remove the background grid.** theme_tufte() has a blank background, but I don't particularly like some of its other defaults. If you'd like to keep another theme but remove the background grid lines, you can add `+ theme(panel.grid.major = element_blank(), panel.grid.minor = element_blank())` to your plot.
- **Adjust text appearance.** While this makes sense within the framework of the grammar of graphics, I find that adjusting text in a `ggplot2` visualization is more complicated than I'd like. There are three things to remember:

1. Appearance of things that *aren't related to your data values,* such as text size and font, are set inside theme(). A theme such as theme_minimal() includes a lot of default look-and-feel settings. You can override any of those defaults inside of a theme() function call.
2. theme() has *element functions.* They set things like font size and color.
3. You can change default appearances with a format like `theme(part.of.graph.to.change = element_function(item = value))`. So, to set the size of the graph title to 24 points, it would be `theme(plot.title = element_text(size = 24))`, and to italicize the x-axis labels it would be `theme(axis.text.x = element_text(face = "italic"))`.

I warned you this part was a bit complicated. The good news is, customizing your colors will be easier than this.

Meanwhile, the best advice I can offer for dealing with ggplot2 text tweaks is to make code snippets for things you know you'll regularly want to change.

6.11 Comment your code

Congratulations! We're now doing code that's complex enough that we may run the risk of not being able to fully understand it if we look back at it three months from now.

If by chance you're thinking, "I'm *sure* I'll remember what this is" ... well, unless you've never misplaced anything *ever,* you may want to reconsider that confidence. It's easier to forget what a block of code does than you might think.

Adding comments - explanatory text that isn't run as code when the script is executed - is an important way to remember exactly what that code is supposed to be doing. But even if you have a photographic memory, there are other reasons to add comments to your scripts:

- You may share your code, whether to work together with a colleague, ask someone to check your work, or publish your script as part of a story or open-source project. Comments help *other people* understand what your code is supposed to be doing.
- If you've got a lengthy script file and want to make a change in part of it, it's easier to find the section of code you need if you've added comments, because then you can search for a keyword within a comment.

Commenting code is the "eat your vegetables" of the programming world. We all know we should do it, but too often we don't. Especially when we're in a rush (which is most of the time), we'll think to ourselves "I'll

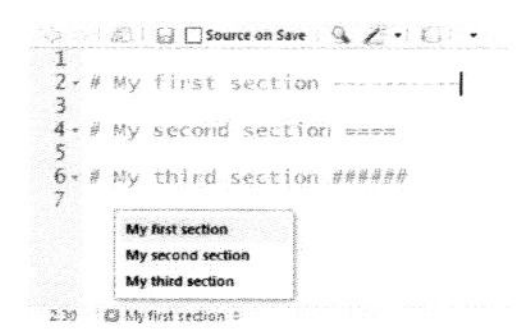

Figure 6.23: Navigating within a script in RStudio.

get to that later" (chances are, we won't) or "I'll remember why I wrote this" (good luck with that). So if you're a new programmer, this would be an excellent time to get into the habit of commenting your code.

In R, anything after a pound sign `#` is a comment. That means it will not run as computer code; it's just there to give you helpful hints.

Here's what some code from the last chapter might look like with comments:

```
# Subset the snowdata data frame to find Chicago winters with
# less than 2 feet or more than 5 feet of snow

chicagoextremes <- snowdata %>% # Store copy of snowdata into chicagoextremes
  filter(Chicago < 24 | Chicago > 60) %>%
  # Keep rows where Chicago is less than 24 or greater than 60
  select(Winter, Chicago) %>% # Select only the Winter and Chicago columns
  arrange(desc(Chicago)) # arrange new data frame by Chicago column in descending order
```

In the real world, you'd never add comments to your code in every line. This kind of over-commenting is usually just for when you're learning. But once you're more advanced, adding some *strategic* comments can help. With a heavily customized ggplot2 visualization, for example, you might want to put a comment line at the top, such as `# bar graph reordered by descending y values`.

Within RStudio, there's another advantage to commenting code: navigation. If you've got a long script file, you can break it up into sections by using special comment formatting and then easily jump from section to section (see Figure 6.23).

A comment that *ends* with *at least* four dashes, equal signs, or pound signs becomes a *section header*. Any of these will work:

```
# My first section ----------

# My second section ====

# My third section ######
```

There are several ways to find and use your section headers. At the bottom left of the console panel, you'll see which section you're in - click on that to get a list of all the section headers, and then click on any of those items to move your cursor into another section.

If you're using this often, you may want to keep an outline panel open within your script console at times. Click the outline button at the top right of the console – it's the one with multiple horizontal lines on it (see Figure 6.24) – and you'll see the full document section outline. Keyboard shortcut Ctrl+Shift+O Windows or Cmd+Shift+O on Mac will open it as well (that's the letter O not a zero). Click that button again or use the same keyboard shortcut to close the outline view.

Figure 6.24: RStudio's outline panel.

6.12 Wrap-up

We've gone over quite a bit in this chapter: multiple ways of visualizing data in R with both base R and ggplot2, creating custom code snippets, and commenting code. You should now have a solid foundation on how to do basic data exploration and visualization in R.

Next up: It's time to wrangle some data files that don't come packaged in just the right format.

6.13 Additional resources

All these links are also available in the booklinks.html file or at https://smach.github.io/R4JournalismBook/booklinks.html.

RStudio's ggplot2 cheat sheet. Download the 2-page PDF at https://www.rstudio.com/wp-content/uploads/2016/11/ggplot2-cheatsheet-2.1.pdf.

Computerworld's searchable, sortable table of ggplot2 commands. I wrote this to make it searchable by task instead of command. Find it at http://cwrld.us/ggplot2ChtSht (there's also a link to some downloadable code snippets).

R Graph Catalog. A searchable catalog of different ggplot2 visualizations with downloadable sample code, maintained in part by RStudio's Jenny Bryan. http://shinyapps.stat.ubc.ca/r-graph-catalog/. Another option is **Top 50 ggplot2 Visualizations – The Master List (With Full R Code).** And while these aren't actually *my* top 50, you'll find some more samples with R code to browse. From r-statistics.co. http://r-statistics.co/Top50-Ggplot2-Visualizations-MasterList-R-Code.html.

ggplot2 graphics companion. Some very nice examples of publication-quality visualizations by type, all with code. From Trafford Data Lab in Trafford, Greater Manchester, UK. http://www.trafforddatalab.io/graphics_companion/.

Base R graphics. Statistician and author Nathan Yau leans toward base R graphics. He did a comparison of base vs. ggplot2 for a number of data visualizations. Although written in March 2016, it should still be relevant: https://flowingdata.com/2016/03/22/comparing-ggplot2-and-r-base-graphics/.

Base R vs. ggplot2. Two advanced R users talk about their respective viewpoints on why base R or ggplot2 is preferable. Maybe you'll be moved by one of their posts. See Why I don't use ggplot2 by Jeff Leek at Johns Hopkins University http://simplystatistics.org/2016/02/11/why-i-dont-use-ggplot2/ and Why I use ggplot by Stack Overflow data scientist David Robinson at http://varianceexplained.org/r/why-I-use-ggplot2/.

Formula-based version of ggplot2. If you like ggplot2's capabilities but find the syntax uncomfortable, the ggformula package on CRAN might be worth exploring. As the package vignette explains, it uses the syntax `gf_plottype(formula, data = mydataframe)`. gf_plottype options are gf_point(), gf_line(), gf_bar(), gf_boxplot(), etc.; while the formula can be something like `yvaluecol ~ xvaluecol`. Install, load, and read the package vignette, `vignette("ggformula",package = "ggformula")`.

esquisse RStudio add-in. This add-in offers a basic drag-and-drop interface for creating simple ggplot2 graphics, including generating and displaying the underlying R code. See more information at https://github.com/dreamRs/esquisse.

Chapter 7

Two or more data sets

When you're first learning to program, sample data sets are usually exactly in the format you need, letting you concentrate on coding fundamentals. In the real world, of course, data is rarely just the way you want it. In this chapter, we'll look at how to deal with data when it's in multiple files and formats.

7.1 Project: Multiple files of U.S. airline on-time data

The U.S. Department of Transportation keeps records of domestic airline flights, including whether each was on time, delayed, or cancelled. Files are broken out by state and month.

There's a lot you can do with this data, such as looking at overall airline performance to and from a specific airport. But for something a bit more dramatic, we'll put together a data frame of air-travel woes during and after 2012's "Superstorm Sandy" in the New York metropolitan area.

This is a good test case for handling data in multiple files. The region's three major airports are in two different states; and since the storm hit on Oct. 29, its travel impacts span two months. I downloaded domestic flight data from the U.S. Department of Transportation for New York and New Jersey in October and November of 2012. Since each month and state has its own file, there are four files in all.

7.2 What we'll cover

- Tapping into some operating-system functions like listing files in a directory
- Adding one table to the bottom of another
- Manipulating data in lists
- Importing all CSV files in a directory at once
- Changing working directories and then easily getting back to your project's main working directory

7.3 Packages needed in this chapter

In addition to having rio installed:

```
pacman::p_load(dplyr, purrr, here)
```

7.4 Add one table to the bottom of another

In Excel, if you had four files, all with the same columns, you might copy one entire spreadsheet and paste its data to the bottom of another until all four were combined into one. In R, you can write a bit of code to do this, so you know all the data has been merged correctly, instead of risking a copy-and-paste error. And, if someone is checking your work or trying to reproduce what you've done, everyone will know that the data is being combined the same way regardless of how many times you run the code.

If you've already downloaded this book's GitHub repo, files will be in your sandydata subdirectory. If you haven't yet, please follow the instructions in Chapter 5.

In R, adding new rows to the bottom of an existing data frame is usually done by *binding rows.* If you had multiple columns you wanted to add to an existing data frame, you could *bind columns.* The base R functions for this are rbind() (for binding rows) and cbind (for binding columns). dplyr's versions are bind_rows() and bind_cols(), and the end result is a data frame. (In base R, rbind() produces a *list of data frames* and then you need to run another command to combine the list into a single data frame. That's why Wickham added bind_rows() to dplyr.)

You could manually type in the names of all the files and then bind them together, which would be mildly tedious but not terrible with only 4 files. But what if there were 14 files? or 40? Here's a better way: Put all the files in their own directory, have R read all the names of the files, and then bind them together.

R has some functions that do operating-system types of commands, such as changing the working directory, listing all the files in a directory, copying files from one directory to another, and so on. Here are a few of them:

- **getwd()** - gets the name of the current directory.
- **setwd("path/to/newdirectory")** changes the working directory
- **list.files()** - lists all files in the current directory. list.files("subdirectory") will list the files in a subdirectory of the current working directory. list.files(pattern = "*.csv") will list all files ending in .csv only (the asterisk is a wildcard, meaning the pattern matches 0 or more of any other characters along with the ones you specify.)
- **file.copy()** - uses the format `file.copy("path/to/file/myfile", "path/to/copy/to", overwrite = TRUE or FALSE)`
- **file.exists("filename")** - checks to see whether a file exists in a certain directory. If no path is given, it checks in the current directory.

Read all the file names and store them in a variable with:

```
sandyfiles <- list.files("sandydata")
```

7.5 What's a list, and how do you operate on one?

As I mentioned in the first chapter, every element in a vector (including a data frame column) has to be the same data type. But what if you need a "collection" of different data types? In R, that's a list. `c(2, 3, "final")` creates a vector of character strings "2", "3", and "final" – since R can't turn "final" into a number, all items are "coerced" into strings. If you want to keep the original data structure of the number 2, the number 3, and the character string final, you'd use the list() function: `list(2, 3, "final")`. However, when you print out a list, it looks a lot different than a vector:

```
mylist <- list(2, 3, "final")
mylist
```

```
## [[1]]
## [1] 2
##
```

```
## [[2]]
## [1] 3
##
## [[3]]
## [1] "final"
```

Here's an absolutely critical point about lists. If you subset a list using *single* brackets, *you get a list back* – even if that item has just a single value, such as the number 2. Run `class(mylist[1])` and you'll see. If you want to get the *value that's inside a list item, use double brackets:*

```
mylist[[1]]
```

```
## [1] 2
```

```
class(mylist[[1]])
```

```
## [1] "numeric"
```

If you're having trouble with this concept, you might try thinking of a list as a *box of envelopes*, with each envelope holding information. mylist[1] gives you the first envelope in the box; but if you want to know what's *inside* the envelope, you have to go another layer deep, with double brackets.

One more point about lists: list items don't have to be a single, basic thing (known as an atomic data type) like one number or one character string. List items can also be, say, a vector. A *data frame* can be one item in a list. (A data frame is actually a specific type of list, one with a specific structure). You can have lists of lists – and if you're using R to parse certain types of data, you will.

But enough theory; let's start working with those data files. First, if you haven't yet, save the *names* of the files in a variable using the list.files() function (I call my variable sandyfiles): `sandyfiles <- list.files("sandydirectory")`.

It would be great if we could just run something like `rio::import(sandyfiles)` to pull in all the files. That won't work, though, because *import() can only take one file at a time.* Fortunately, there are R functions designed to run a function on multiple items at a time, both in base R and external packages.

An easy way to do this used to be with a Wickham-authored package called plyr. Unfortunately, plyr is an older package and isn't included in the tidyverse. Instead, the tidyverse has transitioned to a newer package called purrr, which is much more powerful but also has a bit more of a learning curve.

There's also the option of base R's lapply() function, which is pretty streamlined but needs an extra step compared with purrr. I'll show you both, since I've still encountered some situations where lapply() comes in handy.

7.5.1 mapping with purrr

purrr has a family of `map` functions that *apply a function to each item in a vector or list.* What's handy about these functions is that you can specify the class of data you'd like returned based on the function name. For example, `map_df()` returns a data frame, `map_chr()` returns a character vector, and `map_int()` returns a vector of integers. Plain `map()` returns a list.

The simple syntax for using purrr map functions is: `map(mydata, myfunction, anyAdditionalArguments)`. This says "For each item in mydata, apply the function myfunction, using any additional function arguments I've specified."

There's also a *formula* type of syntax for map, which you can use for more complicated applications. I'll show examples of that in Chapter 15.

For now, let's apply rio::import to each item in sandyfiles, and store the results in sandydf.

First, set your working directory to sandyfiles:

```
setwd("sandydata")
```

Then run map_df() on sandyfiles to apply rio::import. Note that in this case, the function name *doesn't* need parentheses after it.

```
sandydf <- map_df(sandyfiles, rio::import)
```

You should now have a data frame called sandydf with 128,738 rows and 40 columns.

By the way, using a pipe, you wouldn't even need to store the file names into an interim variable sandyfiles. Instead, you could pipe the results of list.files() into the map_df command. Since your working directory should now be sandydata where the files are stored, you'd change list.files("sandydata") (which specifed the sandydata subdirectory of your current working directory) to list.files() (to list files in the current directory)

```
sandydf <- list.files() %>%
  map_df(rio::import)
```

You can now re-set your working directory back to the sandydata parent directory with `setwd("..")`. Those who are familiar with a Unix command line will recognize ".." as being "parent directory one level up".

7.6 lapply

Base R's `lapply()` turns a function that normally runs on one item into a function that can operate on multiple items – as long as those items are in either a vector or a list. The syntax is `lapply(mylist, myfunction)`.

Give this a try. Once again, set your working directory to where your data files are stored using setwd(): `setwd("sandydata")`. If you don't do this, R will look for the files in your main project directory and won't find them. Run

```
sandydata <- lapply(sandyfiles, rio::import)
```

and you will have imported each file *into a list called sandydata.* Each element of that 4-item sandydata list will be a data frame, as you can see if you run `summary(sandydata)` .

dplyr's bind_rows() function can then combine each data frame in the list into one, single data frame:

```
sandydf2 <- bind_rows(sandydata)
```

Using %>% piping, this can be combined into a single statement.

```
sandydf2 <- lapply(sandyfiles, rio::import) %>%
  bind_rows()
```

Remember, this syntax takes the results of the first line and "pipes" it into the second. Now, this doesn't only save a little typing; it also saves memory, since it's not saving a sandydata list that we created above. With a data set of more than 128,000 rows and 40 columns, it's more efficient not to save two copies of the same data.

This is also a fairly elegant way of importing multiple files into R.

It would be useful to check whether this new data frame is the same as the first data frame, using the `identical()` function:

```
identical(sandydf, sandydf2)
```

```
## [1] TRUE
```

They are indeed the same. Now, though, there are two copies of the data, which is one more than we need. As discussed in Chapter 4, you can delete objects from your workspace with the rm() function:

```
rm(sandyfiles, sandydf2)
```

Take a look at the data with `str(sandydf)` and you'll probably see that not all the column names are crystal-clear about what they're measuring.

To make sure you understand what each column is measuring, you can see an explanation of the fields (sometimes referred to as a *schema*) at http://www.transtats.bts.gov/Fields.asp?table_id=236. There are also more definitions if you click on the Table Profile link.

Right now this data includes flights to and from every airport in New York State and New Jersey. I'd like this project to look specifically at flights only at the New York City metro area's three major airports: JFK, LaGuardia, and Newark. To see what airports are in the data set and how they're listed (Are there full airport names? Codes such as JFK and LGA? Something else?), I'd like to get the *unique values* in the aiport columns. R has a simple, intuitive unique() function. I'll look first at unique values in the ORIGIN column, which is the origin airport:

```
unique(sandydf$ORIGIN)
```

```
##  [1] "LAX" "EWR" "DFW" "MIA" "SEA" "MCO" "PBI" "RSW" "SJU" "TPA" "BOS"
## [12] "FLL" "ATL" "MSP" "DTW" "SLC" "CLT" "IND" "BDL" "RDU" "PWM" "DAY"
## [23] "PIT" "STL" "CMH" "MCI" "BNA" "ALB" "XNA" "GSO" "JAX" "BUF" "IAD"
## [34] "SDF" "CVG" "ROC" "SAV" "BWI" "DCA" "MKE" "BTV" "ORF" "OMA" "CHS"
## [45] "OKC" "TUL" "GRR" "RIC" "MEM" "GSP" "MYR" "MHT" "CAE" "MSN" "PVD"
## [56] "DSM" "TYS" "AVL" "ORD" "PHX" "SFO" "SAN" "DEN" "LAS" "IAH" "SNA"
## [67] "PDX" "AUS" "CLE" "MSY" "HNL" "SAT" "MDW" "HOU" "PHL" "BQN" "STT"
## [78] "SBN" "SYR" "TTN" "JFK" "LGA" "LGB" "OAK" "PSE" "SJC" "SMF" "SRQ"
## [89] "SWF" "BUR" "HPN" "ELM" "CAK" "ART" "CRW" "ISP" "BHM" "ACK" "MVY"
```

(Reminder: Case matters. `Unique(sandydf$Origin)` will not generate any data).

So, we can see that the ORIGIN column is using common airport codes. I happen to know that JFK is for Kennedy Airport, LGA is for LaGuardia, and EWR is Newark; and if you fly at all, you probably know the code for your local airport(s). (If you don't, clicking on ORIGIN at that Bureau of Transportation link above will give you a list of codes and airports; or, it's easy enough to search on Google.)

Exercise 1: Try to create a data frame named `ny` containing only flights to and from JFK, LGA, and EWR. The answer is at the end of this chapter.

Even if you know how to do this, you may want to read my answer code, because I will show you a shortcut so you don't have to repeat a column name three times.

A good next step would be scanning the column headers to come up with questions you might like to "ask" this data. There's basic flight information including departure and arrival airports, times, and carriers. Then there are fields measuring things like departure delays, arrival delays, cancellations, and diversions. There are also columns for various *causes* of delays or cancellations. We'll get to analyzing some of those in Chapter 8.

Finally: If you're still in the sandydata subdirectory, don't forget to set your working directory back to the project's main directory with `setwd("..")`. Or ...

7.7 here() you are!

Keeping track of your project's home directory is important if you switch directories during your work. While `setwd("..")` works to go back "home" after you've changed to a subdirectory, things can get more complicated if you've got multiple levels of subdirectories, or even just multiple subdirectories.

It can be tempting to simply hard code a directory path, such as `setwd("C:/Sharon/Documents/MyRProject")`. However, that means your code won't run on any other system that doesn't have the *exact same directory structure.*

The here package was designed to solve this problem. Its here() function returns the value of a project's top-level directory. So, wherever you've navigated to within an RStudio project's directory structure, `setwd(here::here())` should bring you back. Run `here::here()` in your console to see what here thinks is your project's working directory.

7.8 Wrap-up

We've covered how to import multiple files in a directory, merge several data files that all have the same column structure, deal with lists in R, and use functions such as purrr's map_df() and base R's lapply(). We also removed multiple items from an R session with rm(), saw all the unique values in a vector with unique(), and checked if two R objects are exactly the same with identical(). If you read this chapter's Exercise 1 answer, you'll also learn about R's handy `%in%` function.

In the next chapter, we'll start analyzing this meaty data with dplyr's group_by() function. And, after that, we'll learn how to easily graph data by groups.

7.9 Exercise 1 Answer

This is one way to create a data frame for flights to and from the three major New York metro airports:

```
# Create a df for flights to and from JFK, LGA, and EWR
library(dplyr) # if dplyr isn't yet loaded in your working session
ny <- sandydf %>%
  filter(ORIGIN == "JFK" | ORIGIN == "LGA" | ORIGIN == "EWR" |
           DEST == "JFK" | DEST == "LGA" | DEST == "EWR")
```

The code above uses dplyr's filter() function to keep only rows where the ORIGIN or DEST columns equal JFK, LGA, or EWR. I find that a bit more elegant than bracket notation `ny1 <- sandydf[sandydf$ORIGIN == "JFK" | sandydf$ORIGIN == "LGA" | sandydf$ORIGIN == "EWR" | sandydf$DEST == "JFK" | sandydf$DEST == "LGA" | sandydf$DEST == "EWR" , ]` , but it's still cumbersome to type `ORIGIN ==` and `DEST ==` three times. Imagine how much less you'd want to do this if you needed to subset by 6 airports, or 20.

My preferred option is to create a vector with all the airport names, and then use R's special `%in%` function. `%in%` finds *which items in a vector are in another vector.* Here's how it works.

Say you've got a vector of states `mystates <- c("New York", "California", "Massachusetts", "Illinois", "Maine", "Texas", "Pennsylvania")` and want to find out which ones were among America's thirteen original colonies. Create another, lookup vector of those original 13 U.S. states with `original13 <- c("Delaware", "Pennsylvania", "New Jersey", "Georgia", "Connecticut", "Massachusetts", "Maryland", "South Carolina", "New Hampshire", "Virginia", "New York", "North Carolina", "Rhode Island ")`. To find out which items in mystates are in the original13 vector. you can start with `mystates %in% original13`.

The result of the code above is not the actual states, but `TRUE FALSE TRUE FALSE FALSE FALSE TRUE`. %in% gives you a *logical vector* showing which items match and which don't. In this case, the first, third, and seventh items in mystates are in original13 and the rest aren't.

You can use that vector of trues and falses to subset your data, extracting items that are TRUE, with:

`` `mystates[mystates %in% original13]` ``

That's the same as `mystates[c(TRUE, FALSE, TRUE, FALSE, FALSE, FALSE,TRUE)]`.

This same logical subsetting works with dplyr's filter:

```
nyairports <- c("JFK", "LGA", "EWR")
ny <- sandydf %>%
  filter(ORIGIN %in% nyairports | DEST %in% nyairports)
```

The first line creates a vector of NY metro airport codes. The next two lines create a new variable called ny from the sandydf data frame, but filtered to include only rows where the ORIGIN column of airport codes is in one of the nyairports codes or the DEST column is one of the nyairports codes.

As with the mystates code above, %in% will also work with bracket notation subsetting the sandy data frame: `ny3 <- sandydf[sandydf$ORIGIN %in% nyairports | sandydf$DEST %in% nyairports,]` (remember that you need a comma after the expression that subsets rows, indicating that you want all the columns. Without a comma, R doesn't know which columns you want.) I definitely prefer dplyr's filter syntax, but either will work.

7.10 Additional resources

R educator Charlotte Wickham (Hadley's sister), who teaches statistics at Oregon State University, posted a tutorial on the purrr package at https://github.com/cwickham/purrr-tutorial. Also see an accompanying video from the 2017 RStudio user conference at https://www.rstudio.com/resources/videos/happy-r-users-purrr-tutorial/.

RStudio's Jenny Bryan posted a Why purrr? explainer at https://jennybc.github.io/purrr-tutorial/bk01_base-functions.html.

Chapter 8

Analyze data by groups

8.1 Project: Airline on-time data analysis (cont.)

There are a *lot* of possible stories in the airline on-time data. We could look at how many flights were cancelled by date, airline, airport, or some combination of those. We could calculate average or median flight delays by airport and airline, or by destination and date. We could examine how many flights were delayed by over an hour during Sandy. Whatever ways you might want to group the data, dplyr's group_by() function will make this type of analysis simple and elegant.

8.2 What we'll cover

- Calculating statistics by group with dplyr's group_by() and summarize() functions
- Using a lookup table
- Understanding missing values
- Graphing counts in a data frame

I suggest we look first at cancellations, a simpler topic than delays because it's binary. A flight is either cancelled or it isn't; while with delays, the *amount* of time delayed is also important.

Let's get to it.

8.3 Packages needed in this chapter

```
pacman::p_load(dplyr, janitor, ggplot2, rio)
```

You'll also need the ny data frame created in the previous chapter.

8.3.1 Counting cancelled flights

How do we find all cancelled flights regardless of airport or date? The schema at http://www.transtats.bts.gov/Fields.asp?table_id=236 says "Cancelled Flight Indicator (1=Yes)" for the CANCELLED column. So, any cancelled flight will have 1 as the value in the CANCELLED column.

We can count cancelled flights by filtering for CANCELLED == 1 and then counting resulting rows. You can count either with base R's nrow() function or dplyr's tally():

```
ny %>%
filter(CANCELLED == 1) %>%
nrow()
```

```
## [1] 8537
```

Base R's `table()` function will give you counts of all groups in a vector:

```
table(ny$CANCELLED)
```

```
##
##      0      1
## 101501   8537
```

- Since each cancelled flight is represented by the number 1 in CANCELLED while non-cancelled flights are 0, you can also add up all the 1s and 0s to get the cancellation total:

```
sum(ny$CANCELLED)
```

```
## [1] 8537
```

- And, the janitor package has a tabyl() function that works much like base R's table() but adds percent calculations automatically:

```
janitor::tabyl(ny$CANCELLED)
```

```
##  ny$CANCELLED      n    percent
##             0 101501 0.92241771
##             1   8537 0.07758229
```

8.3.2 group_by()

Counting cancelled flights *by date* or *by date and airport* requires two steps: 1) *group data by those categories*, and 2) then do your analysis. dplyr makes this easy. Once you group a data frame with dplyr's group_by() function, *everything else you do afterwards is done within each group.* If you run the sum() function after a group_by(), for example, you'll see a sum for each group. If you ask for a median, you'll get the median by group. If you count rows, you'll get the count by group. And so on. If you count cancellations *without* grouping, you'll get the *total* number of cancellations *in the entire data frame* – *all* dates, *all* airports, etc.

One more dplyr point before we get started: You can't simply run a function like sum() or median() after group_by(). Instead, you need to tell dplyr whether you want either to *summarize* the results – that is, get a *new* data frame with each group's name and summarized results – or *mutate* the existing data frame by adding a new column with results.

Let's first take a look at summarizing the data frame with dplyr's summarize() or summarise() (both the British and American spellings work).

The syntax for summarize using a pipe:

```
newdf <- mydf %>%
  group_by(someColumn) %>%
  summarize(
    NewColumnName = someFunction()
    )
```

That creates a new data frame called newdf. NewColumnName contains the result of someFunction() applied to each group based on someColumn.

Counting cancelled flights in the NY/NJ flight data by group may make this syntax easier to understand. Let's take a look at how to count cancellations by date using group_by(). I'll show you a shortcut for this after, but I want to use this simple counting example to demonstrate summarize().

While base R has the nrow() function to count rows in a data frame, dplyr has an n() function that counts *within summarise()* and can be used with %>% pipes.

Code for this counting cancellations by date is below. Make sure you understand how it works, because similar structures will be helpful for many other projects and analyses:

```
sandyCounts <-  ny %>%
  filter(CANCELLED == 1) %>%
  group_by(FL_DATE) %>%
  summarize(
    Total = n()
  )
```

This code filters the data frame to keep only cancelled flights, groups the data frame by the FL_DATE column, and then creates a summary column called Total with the count of each group. Reminder: You need to use either the summarize() function, which creates a new data frame with *just the summary information*, or mutate(), which keeps the entire data frame and *adds a new column with the summary information.*

This syntax will NOT work:

```
sandyCounts <-  ny %>%
  filter(CANCELLED == 1) %>%
  group_by(FL_DATE) %>%
    Total = n()
```

But this will:

```
sandyCounts2 <-  ny %>%
  filter(CANCELLED == 1) %>%
  group_by(FL_DATE) %>%
  mutate(
    Total = n()
  )
```

mutate doesn't make much sense here, since you end up with a new column in this 3,255-row data frame that repeats summary information on a lot of rows. However, mutate() can make sense for other tasks.

As promised, though, there is a shortcut specifically for counting by groups. That's such a common task that dplyr has a specific shortcut for it: count(), with the format `count(mydf, mygroup)`. Here's the shortcut:

```
sandyCounts2 <- ny %>%
  filter(CANCELLED == 1) %>%
  count(FL_DATE)
```

That says "Filter the ny data frame for rows where CANCELLED equals 1, then count the number of rows for each value in the FL_DATE column." The only difference here is that the count column will be automatically named "n", but obviously you can change that.

Look at the first few rows of the new sandyCounts data frame:

```
head(sandyCounts)
```

```
## # A tibble: 6 x 2
##   FL_DATE    Total
##   <chr>      <int>
## 1 2012-10-01    10
```

```
## 2 2012-10-02    30
## 3 2012-10-03    90
## 4 2012-10-04    32
## 5 2012-10-05    18
## 6 2012-10-06     8
```

and you can see one summary for each row, with each row being a date – the category that was group_by'ed.

To summarize this data by more than one group, just add additional columns to the group_by() function. For example, to count cancellations by date *and airline*, group_by the FL_DATE and CARRIER columns:

```
sandyCounts_dateAndAirline <-   ny %>%
  filter(CANCELLED == 1) %>%
  group_by(FL_DATE, CARRIER) %>%
  summarize(
    Total = n()
  )
```

You can do the same with `count(FL_DATE, CARRIER)` instead of lines 3-6 above.

If the question is "Which airlines were most able to keep flights operating?," only looking at the highest number of cancellations won't tell you the whole story, though. One airline may have had 20 cancellations out of 50 flights, while another may have had 30 out of 200. Calculating *percentage* of cancellations by date and carrier would also be useful.

Here, it would be helpful to think through the steps you need before you start coding.

1. Filter the full New York City data set, keeping only rows where the flight date is between Oct. 27, 2012 and Nov. 3, 2012.
2. Group by flight date, carrier, and whether the flight was cancelled or not.
3. Count number of cancellations and total number of rows.
4. Calculate percent cancelled by dividing cancellations by total number of flights.

One of the nice things about summarize is that you can create more than one column at a time. And, the data in your first summary column is immediately available for additional calculations. We know how to count the total number of rows with n(). How do we count total number of rows where cancellations equal 1? In this case, because each cancellation has a value of 1, getting the *sum* of CANCELLED will also give us the *number* of cancellations.

To look specifically at flights around the storm, I'll zero in on dates between Oct. 27, 2012 and Nov. 3, 2012. The date formats for now are character strings "yyyy-mm-dd", such as "2012-10-27". filter() can extract just those rows where the FL_DATE column is between 2012-10-27 and 2012-11-03 inclusive. Although items in the FL_DATE column are character strings and not R date objects, character strings can be sorted alphabetically.That means greater than and less than operations work with character strings, too. (If the character strings were in "Oct. 27, 2012" format, that wouldn't work and I'd have to turn them into date objects. Otherwise, alphabetically "Nov" comes before "Oct". We'll get to date objects in a future chapter.)

This code works to calculate number of cancellations and percent flights cancelled:

```
sandyCancellationsPcts <- ny %>%
  filter(FL_DATE >= "2012-10-27", FL_DATE <= "2012-11-03") %>%
  group_by (FL_DATE, CARRIER) %>%
  summarize(
     TotalCancelled = sum(CANCELLED),
     TotalFlights = n(),
     PercentCancelled = (TotalCancelled / TotalFlights) * 100
  )
```

Reminder: If you want *all* conditions in the filter, you don't need to specify AND because that's assumed. You only have to explicitly specify when it's *or*.

Run `View(sandyCancellationsPcts)` to look at this table. If the multiple digits after the decimal point annoy you (as they do me), you can add a round() function to round the results of the PercentCancelled column. round(mynumber, 1) will round to 1 decimal place. You can use `PercentCancelled = round((TotalCancelled / TotalFlights) * 100, 1)` in the second-to-last line of the code above to round off the percentages.

8.4 Lookup tables

Since I don't happen to know all the airline codes by heart, seeing the carriers' two-letter abbreviations isn't always helpful. Yes, I can figure out AA is American Airlines and DL is Delta, but what's B6? MQ? There's a lookup table at the Bureau of Transportation Statistics website but it's got some duplicate codes in it. I created a Carrier lookup table for this chapter based on common airline codes that were in use during 2012. You can import that from the files you downloaded from this book's GitHub repository with

```
carriers <- rio::import("data/carriers.csv")
```

There are a few ways to "look up" the airline name by code and add it to the sandyCancellations_percents data frame. I want to show you a couple of useful techniques.

The first way to merge the data sources is a classic "join" of the two data frames. "Joins" are common in the world of databases. If you've ever worked with SQL databases, for example, you're likely familiar with terms like inner and left joins. Basically, a join looks for a common column in each data frame, and then merges the two tables based on that common column, or "key.""

In this case, we've got a column with airline codes in sandyCancellations_percents called CARRIER and a column with airline codes in the carriers look-up data frame called Codes. We want each row in the sandyCancellations_percents data frame to be copied into the new merged data frame, along with a new column with airline names. We don't need to keep all the rows in the second data frame, though, since there will be airlines in the complete list of global airlines that don't necessarily fly into and out of New York.

I like base R's merge() function for this, because I don't have to remember whether I want an inner join or left join. Instead, the first data frame is x and the second data frame is y. I tell it "key" column names in each data frame with `by.x` and `by.y`, and whether I want to keep all rows in x and y with `all.x` and `all.y`. Here, by.x is "CARRIER" (name of the key column in the first data frame), by.y is "Code" (name of the key column in the second data frame), all.x is TRUE and all.y is FALSE.

```
sandyCancelled1 <- merge(sandyCancellationsPcts, carriers, by.x = "CARRIER",
                         by.y = "Code", all.x = TRUE, all.y = FALSE)
```

dplyr has join functions such as inner_join() - keeping only rows where there's a match in both tables – and left_join – keeping all rows in the left (first) table and adding matches when available. The syntax for what we want is:

```
sandyCancelled <- left_join(sandyCancellationsPcts, carriers, by = c("CARRIER" = "Code"))
```

The by syntax here uses a vector with the first and second data frames' joining columns. (For those who know SQL and are comfortable with that format, that's the equivalent of the SQL command `SELECT * FROM sandyCancellations LEFT JOIN carriers ON x.CARRIER = y.Code`.)

Valuable tip from a Hadley Wickham workshop: When doing a left join, it's also useful to do an anti_join(), which only returns rows where there *isn't* a match.

```
anti_join(sandyCancellationsPcts, carriers, by = c("CARRIER" = "Code")) %>%
  nrow()
```

```
## [1] 0
```

We can see that there aren't any rows without a match.

Another technique for using a lookup table comes from Hadley Wickham's Advanced R website and works for looking up one column of information only. It uses *named vectors.* Here's how it works.

If you create a new vector called airlines from the carriers$Airline column and ask for its names(), you'll see that it doesn't have any names:

```
airlines <- carriers$Airline
names(airlines)
```

```
## NULL
```

But you can assign each item a name simply by setting the values of `names(airlines)`. For example, you can assign the carrier codes to be the names of the carrier airlines:

```
names(airlines) <- carriers$Code
head(names(airlines))
```

```
## [1] "02Q" "04Q" "05Q" "06Q" "07Q" "09Q"
```

Other computer languages often call these "key-value pairs". What this means is the name serves as a "key" to "unlock" the associated value. In this case, the basic format `airlines["key"]` will give you the associated value. I say basic, because this syntax returns a *named* value, such as:

```
airlines["WN"]
```

```
##                          WN
## "Southwest Airlines Co. "
```

If you'd like to be sure that you're *only* getting the value and not a value plus a name, you want to remove the name. You can do that with the unname() function to remove the key's name:

```
unname(airlines["WN"])
```

```
## [1] "Southwest Airlines Co. "
```

Finally, if you'd like to get rid of that trailing white space in the value, use R's trimws() function, which removes white spaces before and after a character string: `trimws(unname(airlines["WN"]))`. I created an RStudio code snippet called key_value that makes it easy for me to use this type of lookup after creating my named vector:

```
snippet key_value
  trimws(unname(${1:myvector}["${2:key}"]))
```

Remember that if you want to create your own RStudio code snippet, the second line should be indented with a *tab*, not spaces, and it can be added to the RStudio snippets file by Tools > Global Options > Code > Edit Snippets (button) or with `usethis::edit_rstudio_snippets()`.

We can now add airline names to the sandyCancelled data frame by using the key-value lookup en masse. Here, I create a new Airline2 column using key-value:

```
sandyCancelled$Airline2 <- airlines[sandyCancelled$CARRIER]
```

`airlines["WN"]` returns one value. `airlines[namedVector]` returns an entire vector of values based on the airlines named-vector lookup table.

The two columns getting airline names by two different techniques should be identical; you can check with

```
identical(sandyCancelled$Airline,sandyCancelled$Airline2)
```

Next, let's look at delays among flights that *weren't* cancelled between Oct. 27 and Nov. 3. First, we'll create a data frame of all those flights where CANCELLED does *not* equal 1 (the "doesn't equal" sign is `!=`) and then run the psych package's describe function on the DEP_DELAY column:

```
sandyFlights <- ny %>%
  filter(CANCELLED != 1, FL_DATE >= "2012-10-27", FL_DATE <= "2012-11-03")
psych::describe(sandyFlights$DEP_DELAY)
```

```
##    vars    n mean    sd median trimmed  mad min max range skew kurtosis
## X1    1 7449 9.54 34.62     -2    1.95 5.93 -20 650   670 5.21    43.06
##     se
## X1 0.4
```

DEP_DELAY is in minutes. I'll pause here to feel some sympathy for the folks who were on that maximum 650-minute delayed flight. Then, I'll wonder what flights were delayed by at least 10 hours, selecting just a few of the 42 columns to make results easier to read:

```
sandyFlights %>%
  filter(DEP_DELAY >= 600) %>%
  select(FL_DATE, CARRIER, ORIGIN, DEST, DEP_DELAY, LATE_AIRCRAFT_DELAY,
         CARRIER_DELAY, WEATHER_DELAY, NAS_DELAY, SECURITY_DELAY)
```

```
##      FL_DATE CARRIER ORIGIN DEST DEP_DELAY LATE_AIRCRAFT_DELAY
## 1 2012-10-27      UA    DEN  EWR       650                 588
##   CARRIER_DELAY WEATHER_DELAY NAS_DELAY SECURITY_DELAY
## 1            51             0         0              0
```

Just that one United Airline flight on Oct. 27 from Denver to Newark. Note: All those columns ending with _DELAY represent different causes for delays that the federal government records. We can use the useful dplyr SELECT capability that lets you choose column names by *patterns* such as starts_with() and ends_with(). I can rewrite the select line from the code above:

```
select(FL_DATE, CARRIER, ORIGIN, DEST, ends_with("_DELAY"))
```

In addition to FL_DATE, CARRIER, ORIGIN, and DEST, that code will select any column that ends with "_DELAY".

8.5 Beware of missing values

What if we'd like to get the median and mean weather delay for all flights leaving NY-area airports between Oct. 27, 2012 and Nov. 3, 2012 by airport? It seems like I could just 1) filter the data set for FL_DATE between those dates where ORIGIN is JFK, LGA, or EWR; 2) group by airport; and 3) summarize by median WEATHER_DELAY. But look what happens when I try that:

```
filter(ny, FL_DATE >= "2012-10-27", FL_DATE <= "2012-11-03",
    ORIGIN %in% c("JFK", "LGA", "EWR")) %>%
  group_by(ORIGIN) %>%
  summarize(
    MedianDelay = median(WEATHER_DELAY),
    AverageDelay = mean(WEATHER_DELAY)
  )
```

```
## # A tibble: 3 x 3
##   ORIGIN MedianDelay AverageDelay
##   <chr>        <dbl>        <dbl>
## 1 EWR             NA           NA
## 2 JFK             NA           NA
## 3 LGA             NA           NA
```

Aside: This code didn't create a new variable to store the results but just prints them to screen. Printing out results to your screen is R's default when you're coding interactively in the console if you don't specify something else, such as storing in a variable or file.

But back to the results, which show NA - not available - for all three airports. Why is that? The summary() function may give us a clue:

```
summary(ny$WEATHER_DELAY)
```

```
##     Min. 1st Qu.  Median    Mean 3rd Qu.    Max.    NA's
##     0.00    0.00    0.00    1.57    0.00  696.00   89766
```

There are missing values in our data set, shown by R's NA.

Some R functions will ignore NAs, others such as summary() will separate them out, but many R functions will not compute a value if there are missing values. Considering R's heritage as a data-analysis platform, this makes sense. If you're conducting an experiment to determine the effectiveness of a new medication, for example, and data isn't available for a few of your subjects, what should you do when calculating results of that drug trial? Eliminating all the missing values may not be the right approach, since people who didn't finish the trial may not be the same as those who did.

Or, looking at it another way: The average value of 5, 9, and NA could theoretically be *anything*, depending on what that missing value should be. *R wants you as the data analyst to explicitly decide whether or not it makes sense to simply toss out missing values or handle them another way, such as imputing (estimating) them or turning them into a numerical 0.*

In this case, the NAs are for cancelled flights – you can't compute a delay for a flight that never took off and landed unless you want to say it was infinity – so it's fine to remove them. However, if we're reporting about this, it's important to specify that these are delays for flights that eventually took off. Travellers' typical flight delay *experience* won't be the same as these median flight delays, since people whose flights were cancelled had substantially longer delays – potentially days – that won't be reflected here.

R functions such as mean(), median(), and sum() use the `na.rm` argument to specify whether NAs should be removed before generating results. You can see this in their help files – look at `?median` for an example. Both default to `na.rm = FALSE`, but we can change that behavior with `na.rm = TRUE`:

```
filter(ny, FL_DATE >= "2012-10-27", FL_DATE <= "2012-11-03",
    ORIGIN %in% c("JFK", "LGA", "EWR")) %>%
  group_by(ORIGIN) %>%
  summarize(
    MedianDelay = median(WEATHER_DELAY, na.rm = TRUE),
    AverageDelay = mean(WEATHER_DELAY, na.rm = TRUE)
  )
```

```
## # A tibble: 3 x 3
##   ORIGIN MedianDelay AverageDelay
##   <chr>        <dbl>        <dbl>
## 1 EWR             0.        0.791
## 2 JFK             0.        0.656
## 3 LGA             0.        5.86
```

An alternative way of writing this code would be to filter out rows where WEATHER_DELAY is NA, using the is.na() function – or, more specifically `!is.na()` for "is not NA":

```
filter(ny, FL_DATE >= "2012-10-27", FL_DATE <= "2012-11-03",
    ORIGIN %in% c("JFK", "LGA", "EWR"), !is.na(WEATHER_DELAY)) %>%
  group_by(ORIGIN) %>%
  summarize(
    MedianDelay = median(WEATHER_DELAY),
```

```
    AverageDelay = mean(WEATHER_DELAY)
  )
```

Of the flights that went out, at least half had no weather delay at all. But it looks like there were a few flights at LaGuardia that were delayed substantially by weather. We can quickly create a histogram of the LaGuardia data, first by creating an lga data frame with flights between Oct. 27 and Nov. 3 that originated there, and then using base R's hist() function to create a quick histogram:

```
lga <- filter(ny, FL_DATE >= "2012-10-27", FL_DATE <= "2012-11-03", ORIGIN == "LGA")
hist(lga$WEATHER_DELAY)
```

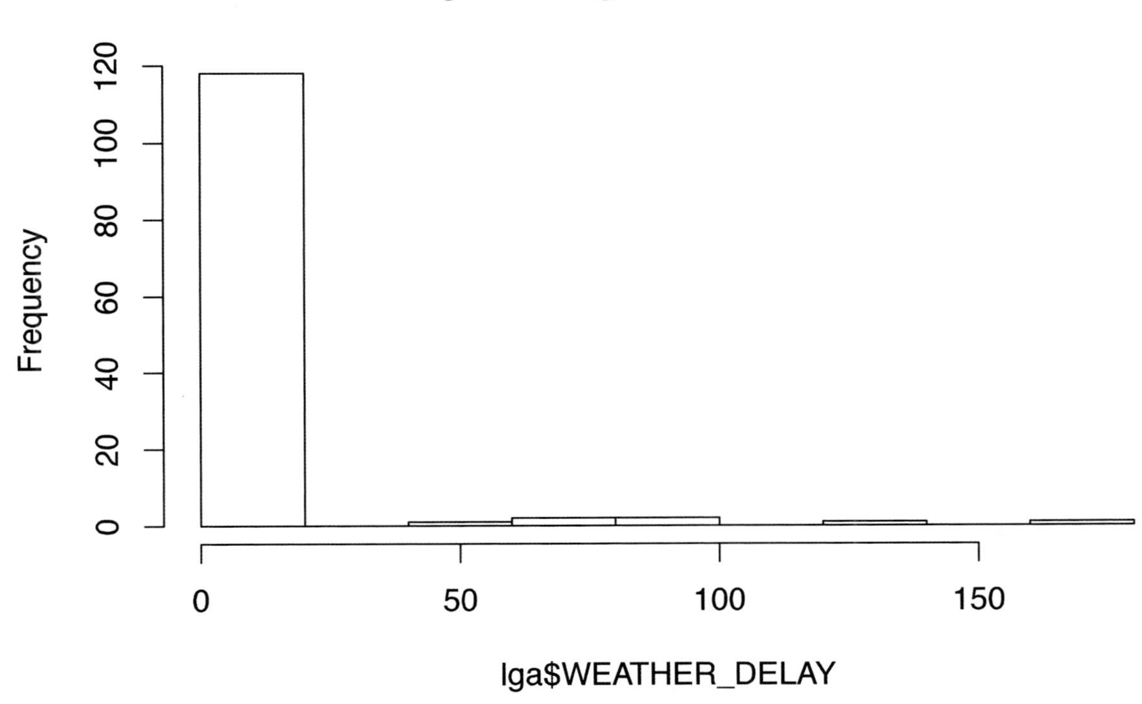

Figure 8.1: Histogram of weather-related delays of flights from LaGuardia during Sandy.

Fortunately, as you can see in Figure 8.1, hist() doesn't choke on NA missing values and just ignores them.

8.5.1 NA vs. NULL

NA is used to represent a missing value. This means the value isn't *available* – it's possible in some circumstances that it could exist but is just unknown. NULL is used when something doesn't exist. If you try to look for a column called Info in the lga data frame by typing `lga$Info` in the console, the response will be NULL – that column doesn't exist, and R is sure it doesn't exist.

NULL can also be used to *delete* a column: `mydf$unwantedColumn <- NULL` removes that unwanted column from a data frame.

8.6 Bar graph of raw data

To generate a bar chart of cancellations, you *don't* need to summarize first.

Remember the default ggplot2 geom_bar() behavior from Chapter 6 when we wanted a simple bar chart of values in a column? geom_bar() was trying to *count items* instead of graphing the *value* of a column, so we needed to add stat = "identity" to geom_bar() in order to graph column values (or use the newer geom_col()). This default behavior works to our advantage now, though, when we have one record per row and want to graph counts.

Graphing two months of cancellations may make for a busy graph with some data that doesn't add to our understanding, so I suggest creating a data frame of just cancellations between Oct. 27 and Nov. 3.

```
sandyCancelled <- ny %>%
  filter(CANCELLED == 1, FL_DATE >= "2012-10-27", FL_DATE <= "2012-11-03")
```

Now look what happens (see Figure 8.2) if you graph with plain geom_bar() and additional arguments:

```
# Make sure you've got ggplot2 loaded with library(ggplot2) or pacman::p_load(ggplot2)
ggplot(sandyCancelled, aes(FL_DATE)) +
 geom_bar()
```

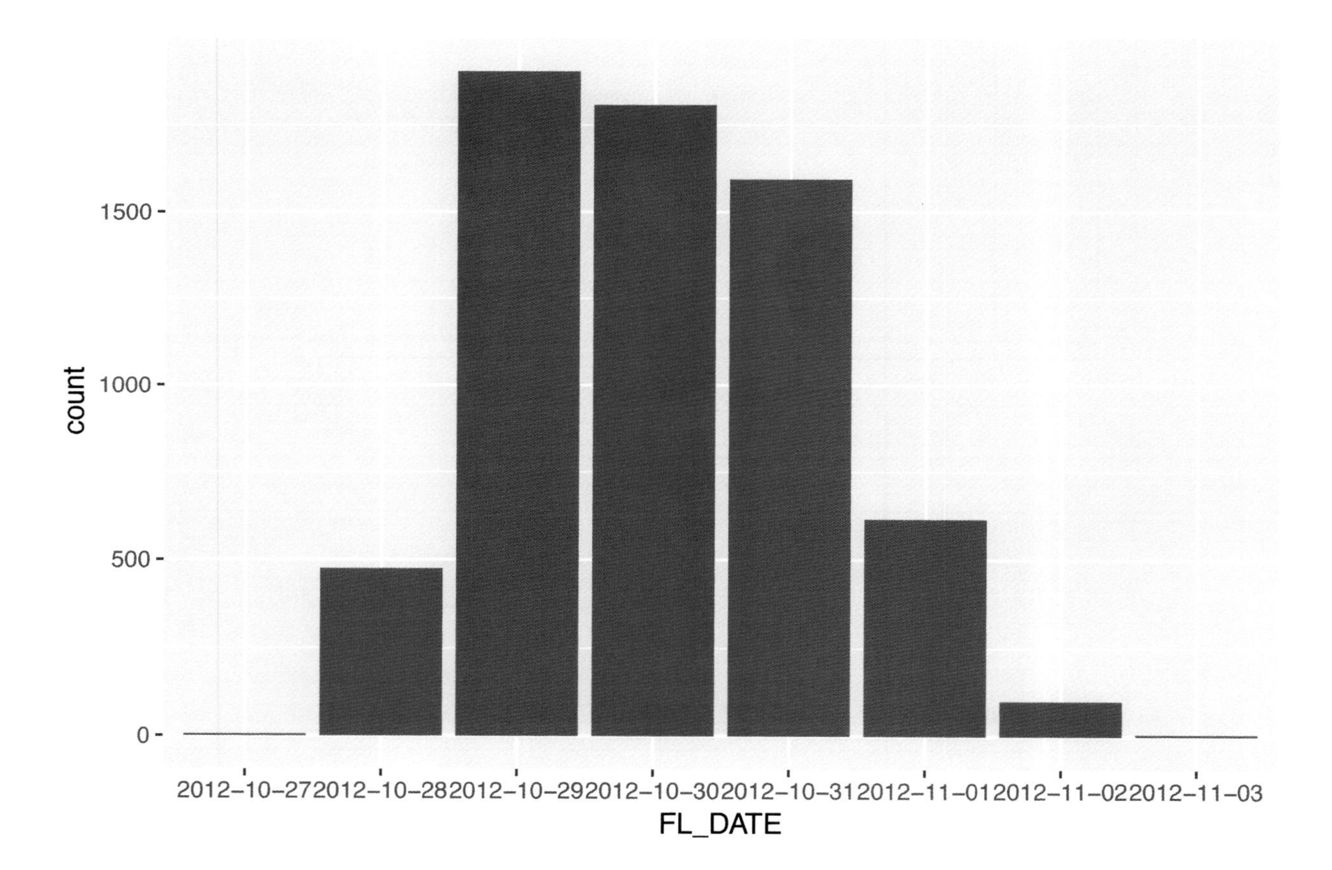

Figure 8.2: Graphing cancelled flights with ggplot2 and geom_bar().

It does seem that Sandy had an impact on cancellations!

ggplot2's geom_bar() default turned out to be useful in this case.

8.7 Wrap up

We used dplyr to summarize by one category and multiple categories and ggplot2 to easily create graphs from a fairly large, raw data set. We also merged data frames by a common column and learned how to use a lookup table (actually, a named vector).

Next up: Handy ggplot2 functions for creating multiple graphs by category.

8.8 Additional resources

For more on using dplyr's join functions, see the two-table vignette with `vignette(package = "dplyr", topic = "two-table")`. dplyr's introductory vignette has good additional explanations for summarize(), mutate(), and group_by(). See it with `vignette(package = "dplyr", topic = "dplyr")`.

There are other ways to summarize by group. Base R has an aggregate() function, the doBy package was designed for statistical calculations by groups, and the data.table package is extremely powerful and fast. I've settled on dplyr for this type of work because I find the syntax intuitive. Many journalists agree.

Syntax, though, is a matter of personal preference. If you'd like to explore alternatives to dplyr, look at the aggregate() help page with `?aggregate`, the doBy introductory vignette at https://cran.r-project.org/web/packages/doBy/vignettes/doBy.pdf, or the yhat blog post about summary stats with data.table at http://blog.yhat.com/posts/fast-summary-statistics-with-data-dot-table.html.

Finally, for very quick summaries by group, the psych package offers describeBy() in addition to describe(). After loading the psych package, see the help page with `?describeBy`. Or, without loading the package, `??describeBy`.

Chapter 9

Graphing by Group

Calculating statistics by group is one useful way to analyze the Sandy flight-delay data. Another helpful technique is *visualizing* data by groups.

9.1 Project: Visualizing airline on-time data

We looked briefly at using ggplot2 with grouped data in Chapter 6 when we created box plots of snowfalls in Boston, New York, and Chicago. But ggplot2 has another interesting way of displaying data by groups: *facets*, which automatically generate a different plot for each category.

9.2 What we'll cover

- ggplot's facet_grid() and facet_wrap()
- color palettes
- packages to customize and expand ggplot2 functionality

9.3 Packages needed in this chapter

```
pacman::p_load(dplyr, janitor, ggplot2, geofacet, RColorBrewer)
```

You'll also need the ny and sandyFlights data frames created in Chapter 7 and used in Chapter 8, as well as rio on your system.

9.4 Facets

Faceting generates separate graphs for each group in a data frame. It will be useful for visualizing the percent of cancelled flights by airport and day. But first, we need to calculate those percentages.

Exercise 2: What was the percent of cancelled flights each day among flights that were supposed to leave during the Sandy time period? Create a data frame called `departing_cancellations` from the ny data. Hint: You first want to filter for flights *leaving* JFK, LGA, and EWR; group by day; summarize the number of cancellations *and* total flights; and then calculate percents. (Answer is at the end of this chapter.)

Once you have this data, you could visualize it with a grouped bar chart (see Figure 9.1) with code such as:

```
ggplot(departing_cancellations, aes(x=FL_DATE, y=PctCancelled, fill=ORIGIN)) +
  geom_col(position="dodge") +
  theme(axis.text.x = element_text(angle = 45, vjust = 1.2, hjust = 1.1))
```

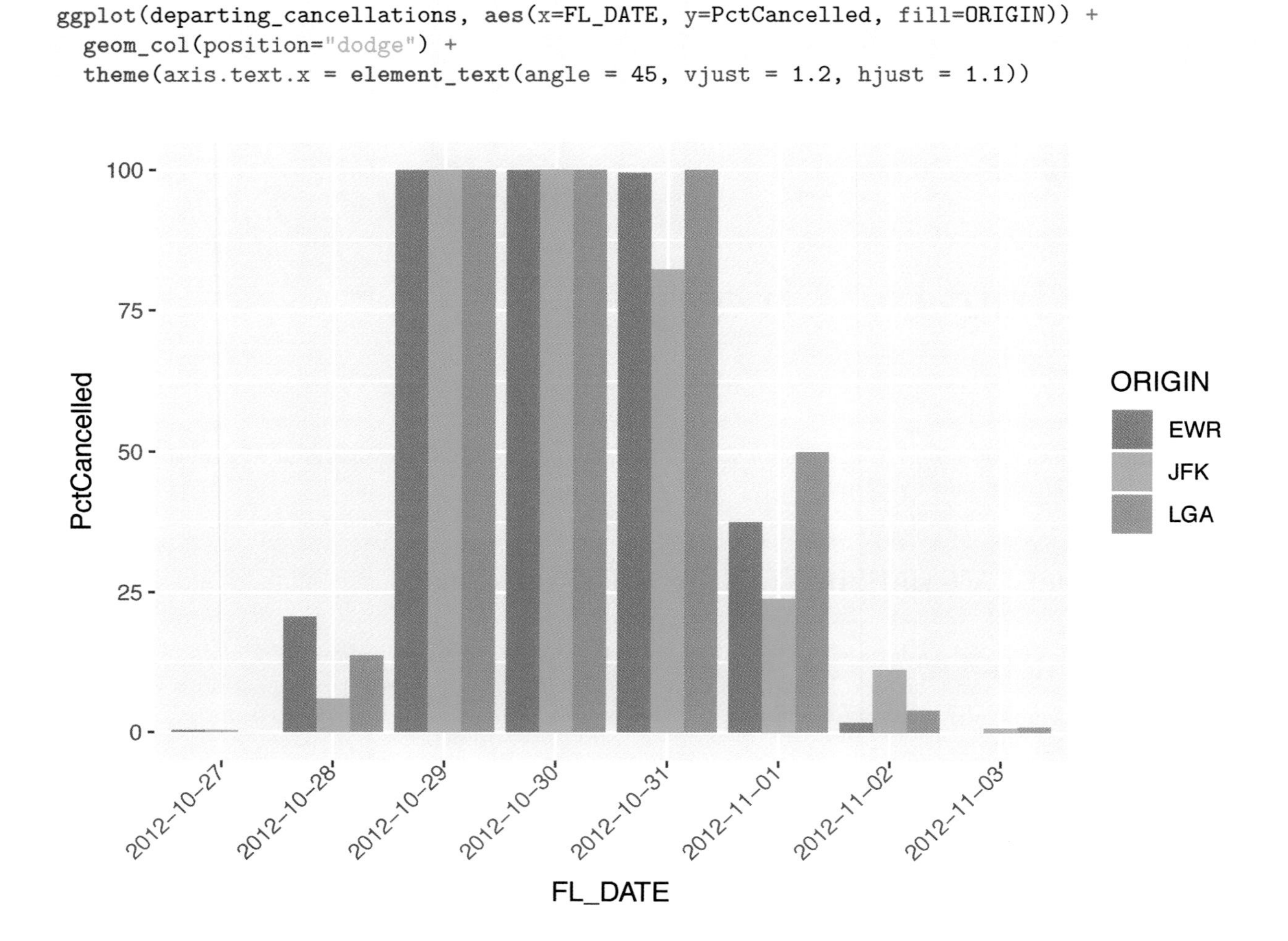

Figure 9.1: A grouped bar chart of cancellations by date and airport.

(Reminder that the last line of code rotates the x-axis text so it's legible.)

But another option is to create side-by-side individual graphs for each airport.

There are a couple of different ways to do facets in ggplot2. With facet_wrap(), you can specify categories, number of rows, and number of columns. facet_grid() is designed for *when all combinations of variables exist in the data* and so it calculates number of rows and columns for you. ggplot2's help files advise that facet_wrap() is generally a better use of screen real estate, but I've had problems running facet_wrap() with bar charts. So, I suggest facet_grid() for this task.

To generate individual graphs by *two* categories, a facet formula looks like this: `facet_grid(category1 ~ category2)`. For splitting by only *one* category, as we want to do here (originating airport), a dot is used in the formula along with the category you're faceting. That syntax looks like `facet_grid(. ~ ORIGIN)` (for a horizontal layout) or `facet_grid(ORIGIN ~ .)` (for vertical). Note that the operator there is a tilde, not a hyphen.

This code below generates a graph for each airport, in a horizontal layout (see Figure 9.2).

```
ggplot(departing_cancellations, aes(x=FL_DATE, y=PctCancelled)) +
  geom_bar(stat="identity") +
```

```
  theme(axis.text.x= element_text(angle=45, hjust = 1.3, vjust = 1.2)) +
  facet_grid(. ~ ORIGIN)
```

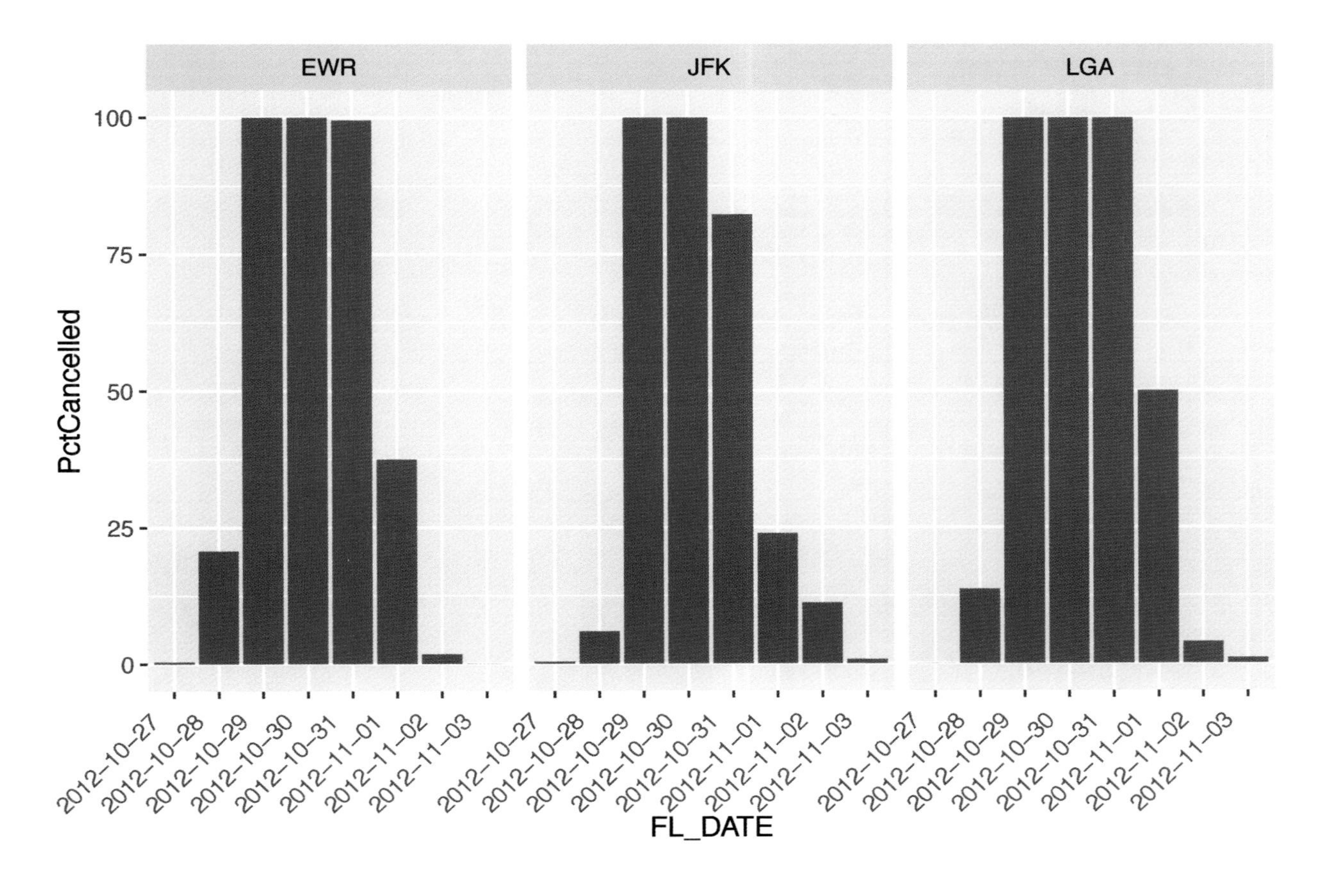

Figure 9.2: ggplot2 facets create one graph for each group.

Once again, I also added a theme statement to rotate x-axis text so it's readable. You can also add fill=ORIGIN to `aes(x=FL_DATE, y=PctCancelled)` if you'd like the graphs to have different colors.

Next question: What did the *raw number of flight cancellations* (not percentages) look like by date, airport, and carrier? This might be what a business writer would look at, to show the *impact of New York cancellations on various airlines' operations*, as opposed to looking at each airport's performance.

Because this question looks at *simple counts* and not *percents*, there's no need to create a separate data frame with calculations. As we saw in the last chapter, ggplot2's bar chart defaults to counts. So, filtering the sandyCancelled data frame for flights originating in the New York metro area will be enough for ggplot2 to graph counts; there's no need to calculate them:

```
departing_cancelled_raw <- filter(sandyCancelled, ORIGIN %in% c("JFK", "LGA", "EWR"))
```

(Syntax note: Because I'm only doing one action here, I didn't use the pipe %>% with dplyr's filter() function. Instead, I used the `function(data, argument)` syntax with filter().)

First, I'm interested in counts of cancellations by date at all three NY-area airports combined, using the departing_cancelled_raw data frame. Flight date is the x axis value, and y defaults to counts in this code:

```
ggplot(departing_cancelled_raw, aes(x=FL_DATE)) +
  geom_bar()
```

Next, I'll look at those counts broken out by airport, with the x-axis text rotated:

```
ggplot(departing_cancelled_raw, aes(x=FL_DATE)) +
  geom_bar() +
   theme(axis.text.x= element_text(angle=45, hjust = 1.3, vjust = 1.2)) +
  facet_grid(. ~ ORIGIN)
```

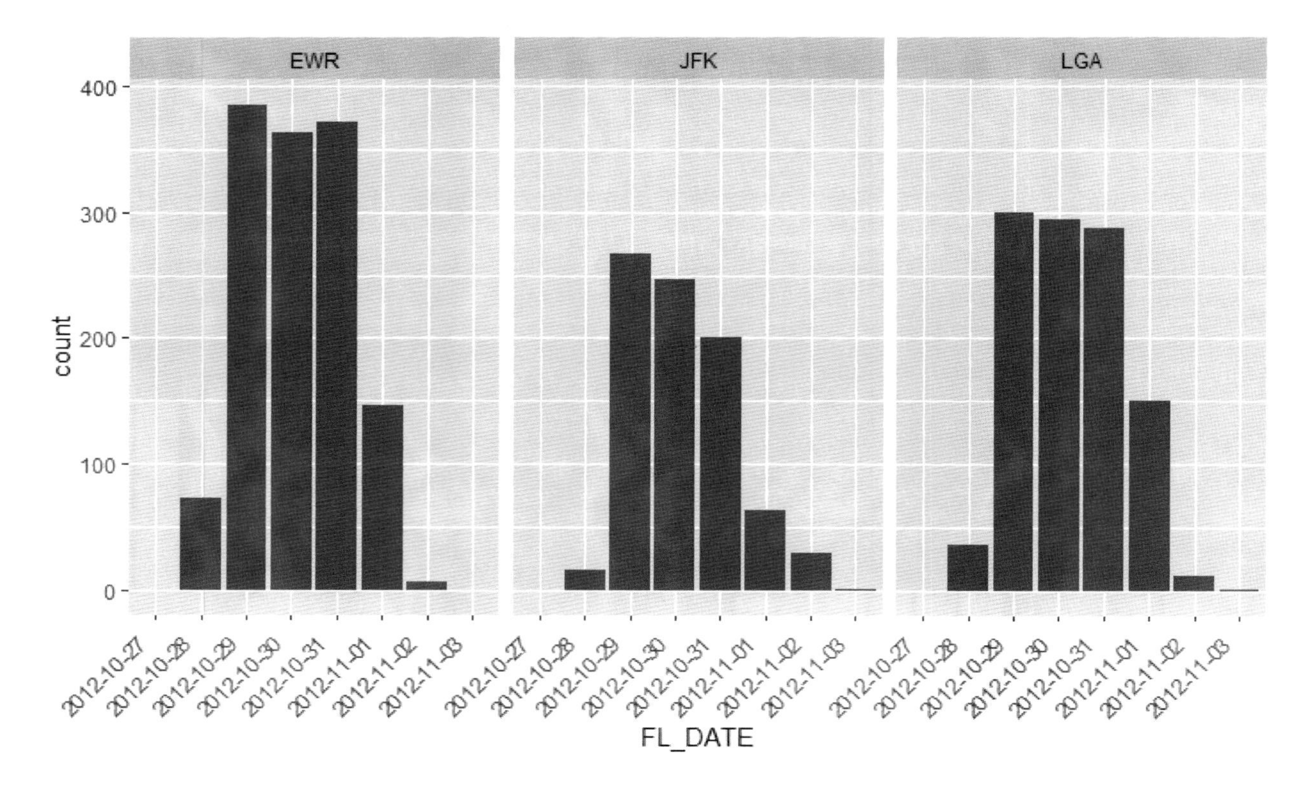

Figure 9.3: Number of cancellations by airport.

If you're familiar with New York airports, you might be surprised to see in Figure 9.3 that Kennedy – the region's busiest – has the lowest totals. However, this data is for domestic flights only, so none of JFK's many international flights are included.

This code visualizes cancellations by airport and destination:

```
ggplot(departing_cancelled_raw, aes(x=FL_DATE)) +
  geom_bar() +
   theme(axis.text.x= element_text(angle=45, hjust = 1.3, vjust = 1.2)) +
  facet_grid(CARRIER ~ ORIGIN)
```

If you run the code and enlarge it in your console, it will be fairly easy to see which airlines were heavily affected and at what airports.

If you were a reporter or transportation policy analyst in Atlanta and wanted to see how that city's air traffic was affected by cancelled (direct) flights originating at NY airports, you could run the same code, but filtering the departing_cancelled_raw data set for destination Atlanta (see Figure 9.4). Instead of creating a new data frame, I can filter() the data set right within the ggplot() function:

```
ggplot(filter(departing_cancelled_raw, DEST == "ATL"), aes(x=FL_DATE)) +
  geom_bar() +
   theme(axis.text.x= element_text(angle=45, hjust = 1.3, vjust = 1.2)) +
  facet_grid(CARRIER ~ ORIGIN)
```

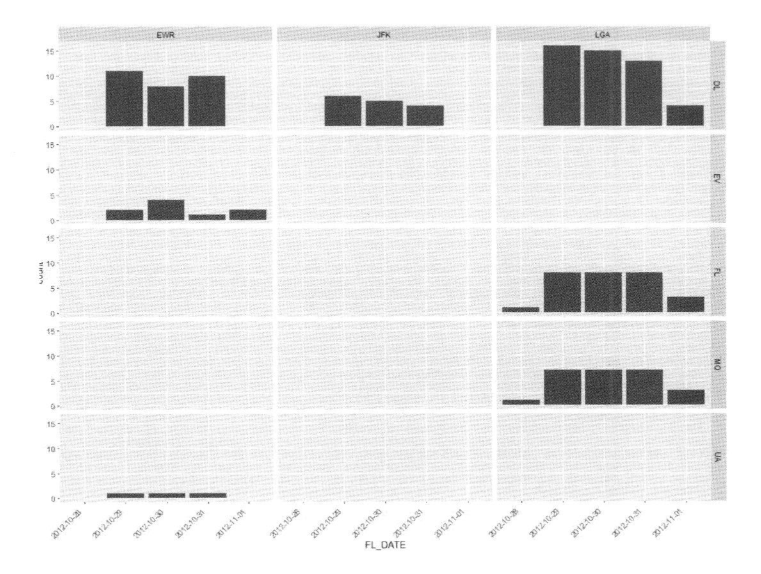

Figure 9.4: Cancellations by airport from the New York metro area to Atlanta.

Similar code could show the effects of Sandy cancellations at other airports.

9.5 Housing prices by state

With only three categories, a grouped bar chart might work just as well as facets. But what if you've got dozens of categories, such as data by state?

The U.S. Federal Housing Finance Agency tracks quarterly mean and median single-family housing prices by state. You can download it from the FHFA data page at https://www.fhfa.gov/DataTools/Downloads/Pages/House-Price-Index-Datasets.aspx or directly at https://www.fhfa.gov/DataTools/Downloads/Documents/HPI/state_statistics_for_download.xls. I downloaded it to the data subdirectory in this book's GitHub repo, so if you've downloaded that, the "state_statistics_for_download.xlsx" file should be in your data subdirectory. If you look at the spreadsheet, you'll see the data starts on row 5; so, when importing, skip the first 4 rows:

```
homeprices <- rio::import("data/state_statistics_for_download.xls", skip = 4)
```

You can take a look at the structure of the file with commands like `str(homeprices)`, `head(homeprices)`, `tail(homeprices)` and `summary(homeprices)`.

In addition, see what time frames are available with `Hmisc::describe(homeprices$ `Year-Quarter`)` or `unique(homeprices$ `Year-Quarter`)`.

It will be helpful to make the column names R-friendly – in other words, remove spaces and hyphens – with the janitor package's clean_names() function:

```
homeprices <- janitor::clean_names(homeprices)
```

The code below gets rid of all the blank rows by only keeping rows where the state column isn't NA:

```
homeprices <- filter(homeprices, !(is.na(state)))
```

Reminder that is.na(state) identifies all the rows where state *is* NA; the `!` sign *negates* that statement, looking for all rows where state is *not* NA.

With a time series – data reported for regular time intervals such as months or quarters – a line chart is a typical visualization choice. For faceted line charts, I'll use facet_wrap, which seems to work better with lines than bars. And, it allows me to set number of columns. The line chart code has to include how to group the data – what group each mini line chart should represent – so add `group = state` to the aesthetic.

The syntax for facet_wrap using one category is `facet_wrap(~category1)`, no dot needed:

```
ggplot(homeprices, aes(x=year_quarter, y=median_price, group = state)) +
  geom_line() +
  facet_wrap(~ state, ncol = 8)
```

To get rid of any scientific notation for housing prices on the y axis, add the general R option `option = scipen(999)`. That tells R to stop using scientific notation this session altogether. Rotating the x-axis text on Figure 9.5 still won't make it readable, so you can get rid of it with the theme() options below (see Figure 9.6).

```
options(scipen = 999)
ggplot(homeprices, aes(x=year_quarter, y=median_price, group = state)) +
  geom_line() +
  facet_wrap(~state, ncol = 12)+
theme(axis.text.x=element_blank(),
        axis.ticks.x=element_blank()
      )
```

Takeaway: If you've got a data set with multiple categories, graphing facets can help give you a better idea of trends within categories as well as how they compare. And it only takes a few minutes to generate several different graphs to explore your data.

9.6 Geofacets

When looking at data by geography such as country or state, another obvious option might be to create a map. In this case, because we're looking at a series of data over time and not just a single data point in each state, a single map might not work (although animating color-coded maps over time might).

However, another compelling alternative is faceted charts like this, but arranged sort of like they might be on a map. With geofacets, you keep the grid and same-sized charts, so, unlike a choropleth map, Texas won't look massively larger than Delaware. But the charts aren't arranged alphabetically.

There's an R package designed to create geofacets with dozens of built-in grids for you to use. Install and load with `install.packages("geofacet")` and `library(geofacet)` or just `pacman::p_load(geofacet)`.

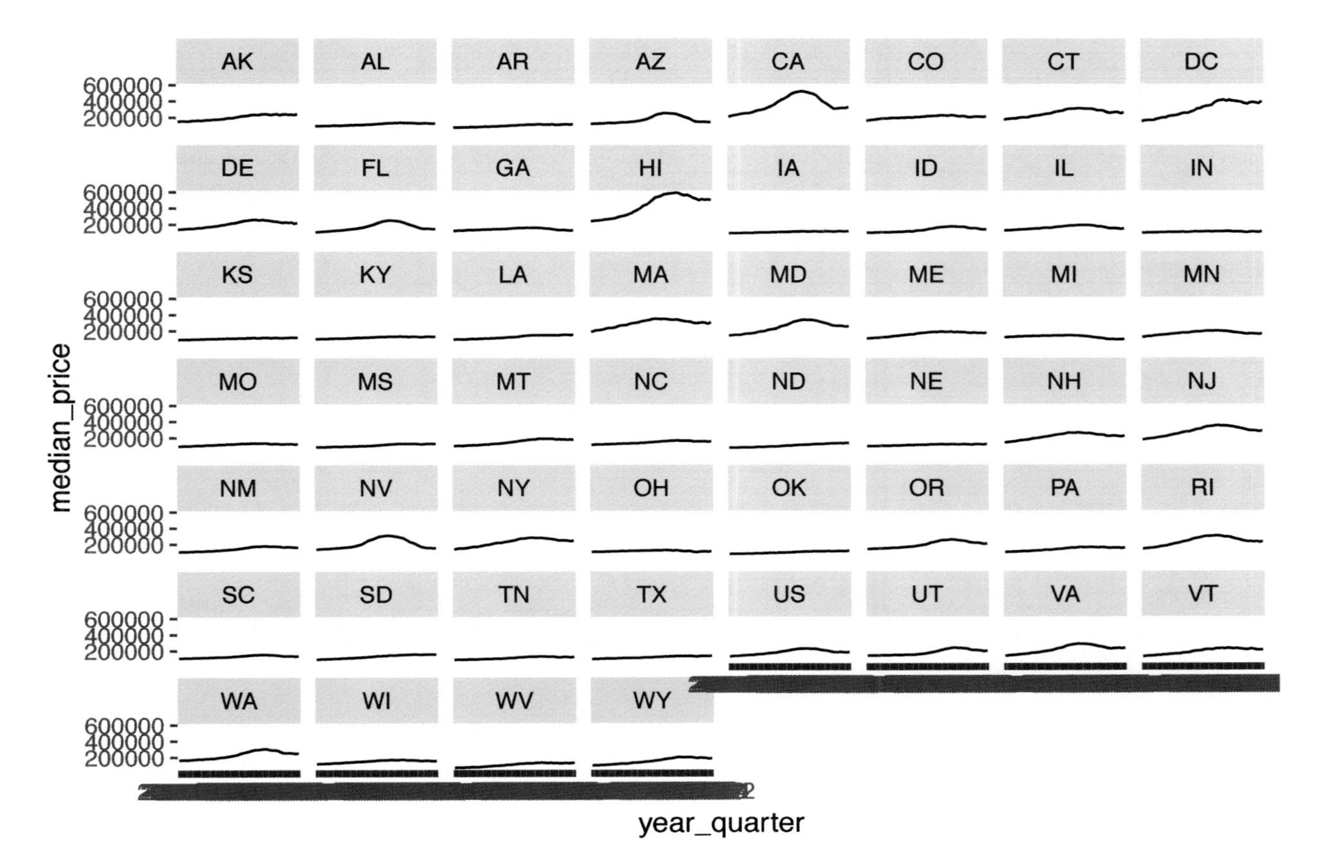

Figure 9.5: Facets of home prices by state.

You can then see all available grids with `get_grid_names()`. Package author Ryan Hafen also included a cool tool in the package to build your own grids for geographies that might not be included, such as city precincts or towns in a county. You can see instructions at https://hafen.github.io/geofacet/.

We don't need to build our own grid arrangement, since geofacet comes with us_state_grid1 and us_state_grid2. Preview a grid with grid_preview(grid_name) such as `grid_preview(us_state_grid2)`. I like that one, since it has Alaska and Hawaii in sort-of geographically correct spots (although way closer to the contiguous 48 states than in reality).

The only change you need to make from the general ggplot2 facet in the previous section is to add `facet_geo(~ state, grid = "us_state_grid2")` instead of the facet_wrap(). I also added a few options to theme() to remove axis text, tick marks, and titles so as not to distract from the visualization:

```
ggplot(homeprices, aes(x=year_quarter, y=median_price, group = state)) +
  geom_line() +
  facet_geo(~ state, grid = "us_state_grid2") +
  theme(axis.text.x= element_blank(),
        axis.ticks.x=element_blank() ,
        axis.ticks.y=element_blank() ,
        axis.text.y=element_blank(),
        axis.title.x=element_blank(),
        axis.title.y=element_blank()
        ) +
  ggtitle("Quarterly Median Home Prices 2000-2010")
```

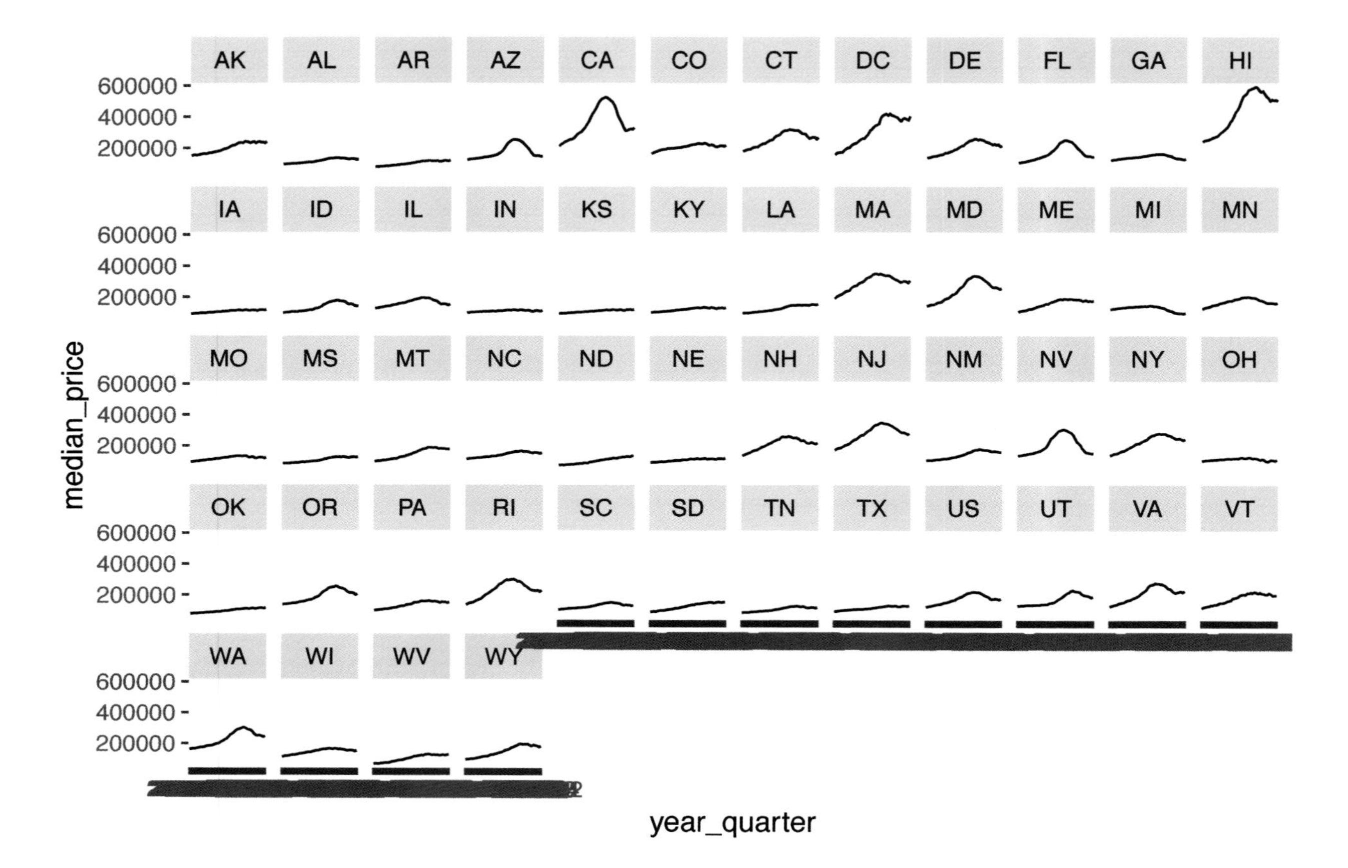

Figure 9.6: Facets after removing x-axis text and tick marks.

There are a couple of advantages to this format, which you can see in Figure 9.7. One, most Americans know roughly where states are on a map of the U.S., compared with trying to find them in an alphabetically ordered grid. Two, when the facet charts are zoomed large enough, you might be able to notice regional trends that would be tough to spot in an alphabetical grid.

9.7 Customizing colors

Let's go back to Sandy data and look at another visualization by group: flights that were *delayed* (not cancelled) taking off in metro NY, compared to the *time in the air* of the flight, grouped by airport. This looks at whether there's a visible relationship between scheduled duration of a flight and how long it was delayed. A scatter plot – geom_point() – will probably work well for this.

With scatter plots, adding a *jitter* functionality helps ensure that you don't have a lot of points with the same value on top of each other, potentially obscuring important trends. If a lot of flights of the same length weren't delayed at all, you'd miss a big cluster that could appear as a single point. *jitter* adds a bit of random noise to the points so they're less likely to be exactly on top of each other.

In the code below, I'm filtering the sandyFlights data set for CANCELLED not equal 1 (in other words, flights that were *not* cancelled) and ORIGIN being one of the three NY airports. The scatter plot will then use AIR_TIME for the x axis and DEP_DELAY for the y axis (See Figure 9.8).

```
delayed_ny <- sandyFlights%>%
  filter(CANCELLED != 1, ORIGIN %in% c("JFK", "LGA", "EWR"))
```

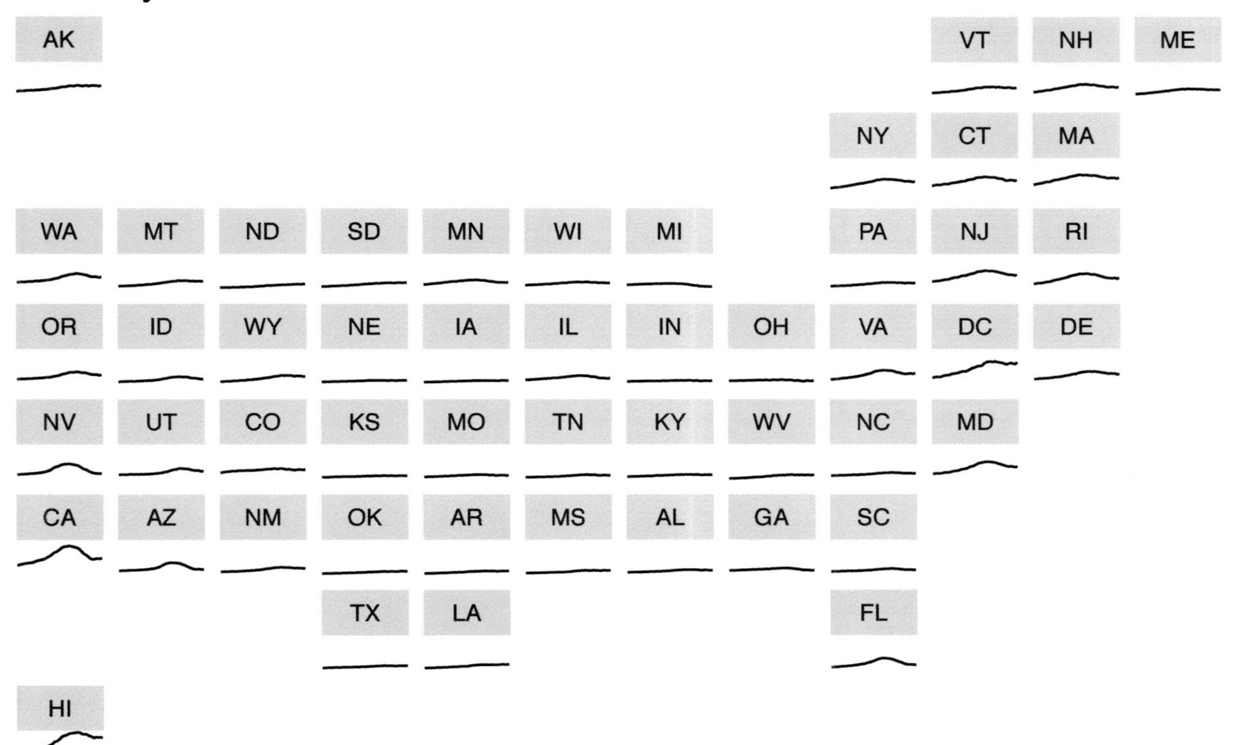

Figure 9.7: Geofacets can be a compelling way to visualize data by region, such as U.S. states.

```
ggplot(delayed_ny, aes(x = AIR_TIME, y = DEP_DELAY, color = ORIGIN)) +
  geom_point(position = "jitter")
```

```
## Warning: Removed 3 rows containing missing values (geom_point).
```

We see that long flights weren't affected by lengthy takeoff delays, although neither were a lot of other flights. Most of the longest departure delays – 300 minutes or more – were from Newark.

In addition, ggplot told us that three rows didn't have values available. You can check to see which ones with this code:

```
filter(delayed_ny, is.na(AIR_TIME))
```

That says "show me all rows in delayed_ny where AIR_TIME is not available."

Base R format would be `delayed_ny[is.na(delayed_ny$AIR_TIME),]`.

Let's use this scatter plot to learn how to customize ggplot2 colors.

You can set your own colors in a ggplot visualization with a `scale_` function: more specifically, either a `scale_fill_` function or `scale_color` function.

Use `scale_fill_` if your visualization type used `fill=` to map colors to categories. Bar charts are one type of visualization that does this (see Figure 9.9), so scale_fill_manual() is the choice:

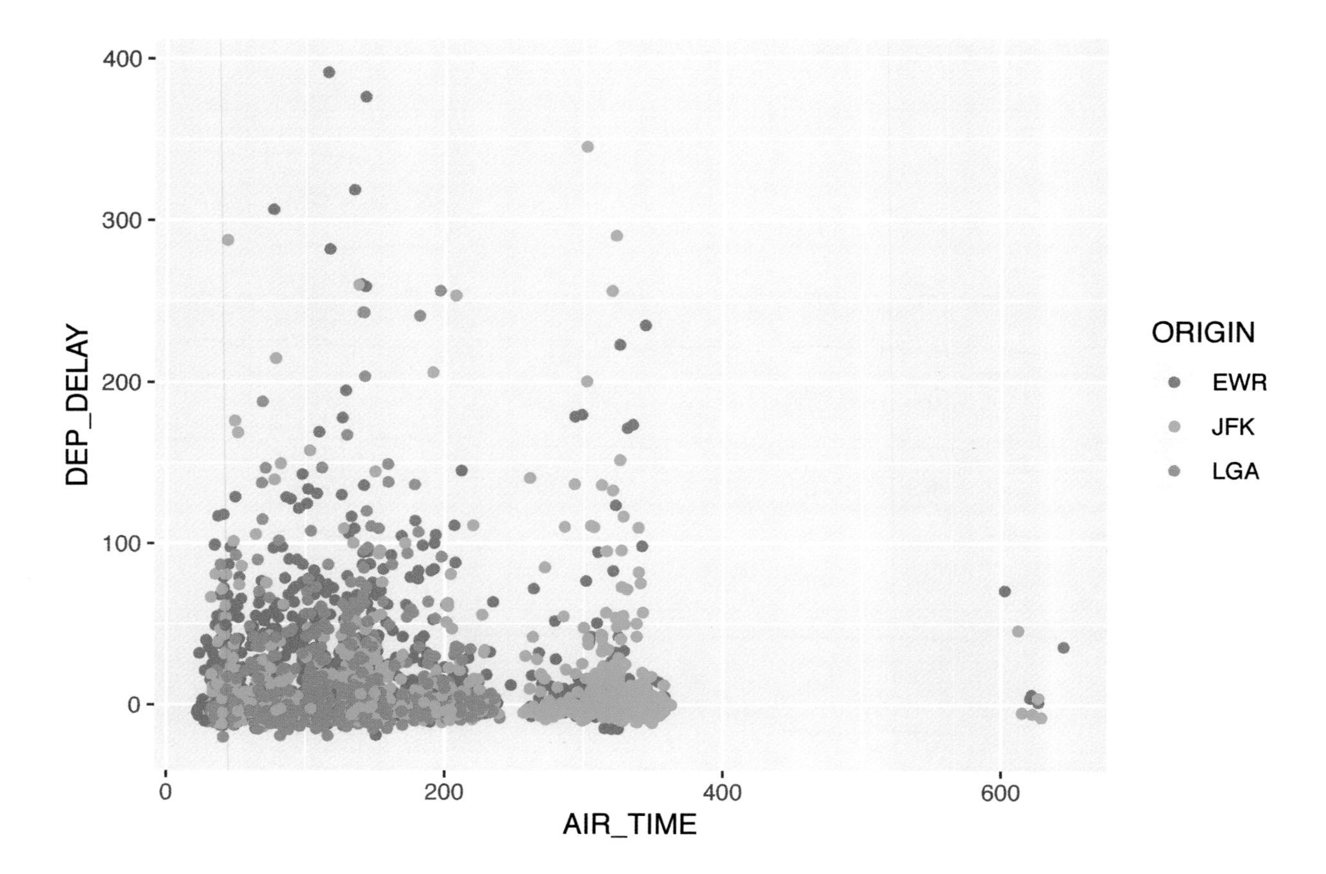

Figure 9.8: A scatter plot looking at flight time versus departure delay.

```
ggplot(departing_cancellations, aes(x=FL_DATE, y=PctCancelled, fill=ORIGIN)) +
  geom_col(position = "dodge") +
  scale_fill_manual(values = c("black", "darkgrey", "white"))
```

Note the syntax: `scale_fill_manual(values = c("color1", "color2", "color3"))`.

In a *scatter plot* (see Figure 9.10), you use `color=` to map colors to categories, so scale_color_manual() is the customizing function of choice:

```
ggplot(delayed_ny, aes(x = AIR_TIME, y = DEP_DELAY, color = ORIGIN)) +
  geom_point(position = "jitter") +
  scale_color_manual(values = c("black", "darkgrey", "white"))
```

```
## Warning: Removed 3 rows containing missing values (geom_point).
```

You can use color names or hex codes for the values in these scale_ functions. scale_fill_grey() and scale_color_grey() create a greyscale palette automatically. Or, you can use pre-defined color palettes.

9.8 Color palettes

The RColorBrewer package is an R interface to the popular ColorBrewer family of palettes, created to help designers choose effective color schemes for maps. Install RColorBrewer with `install.packages`

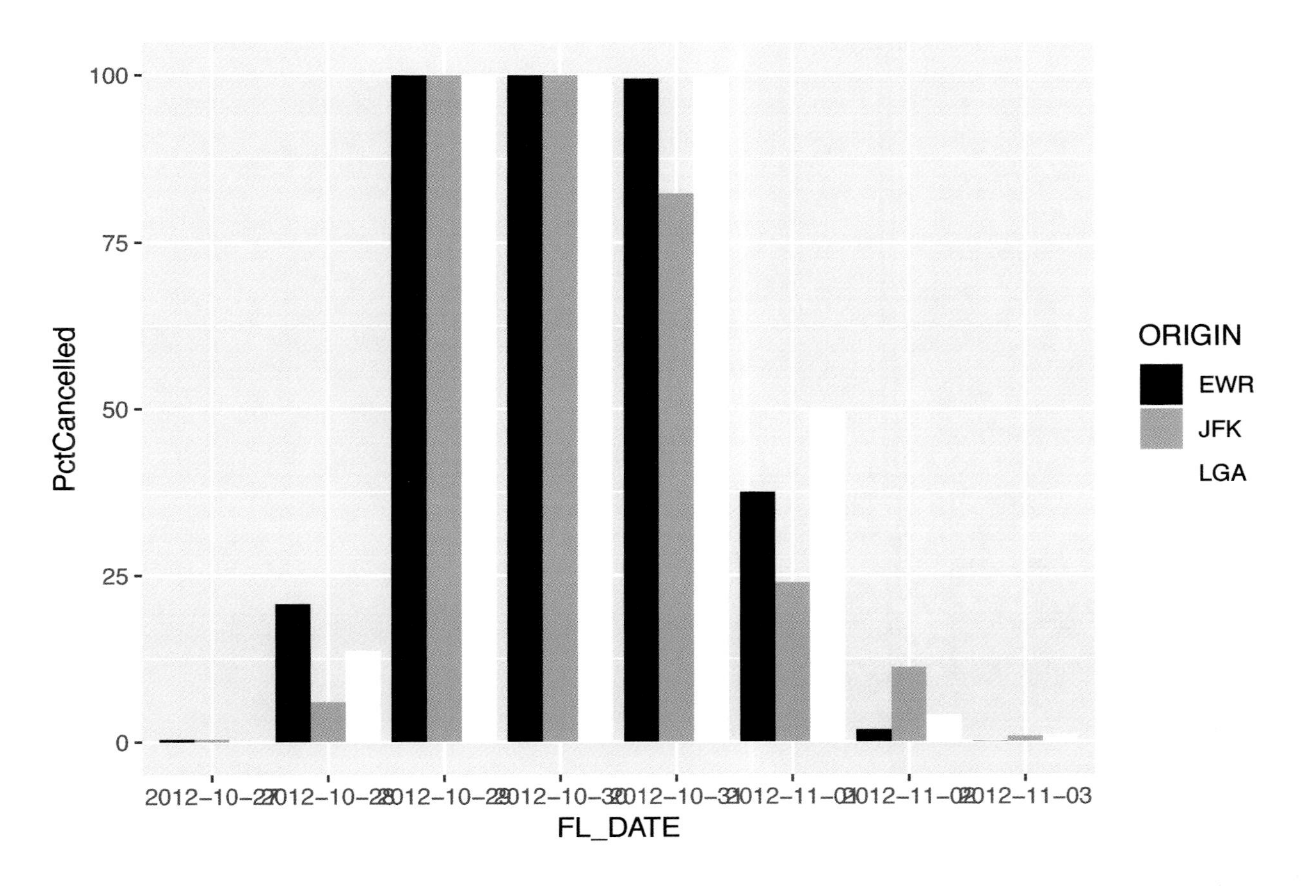

Figure 9.9: Customizing colors with scale_fill_manual.

`("RColorBrewer")` or `pacman::p_load(RColorBrewer)` (if you didn't run the p_load() command at the start of this chapter).

RColorBrewer lets you generate ColorBrewer palettes for *any* R visualization. Run `?RColorBrewer` for more information.

These palettes are also built into ggplot2 with scale_fill_brewer() and scale_color_brewer() functions. Inside those functions, you specify *which* Color Brewer palette you want to use. For the grouped bar chart, one option would be Dark2 with 3 colors. Code for that: `scale_fill_brewer(palette="Dark2")` (no need to specify the number of colors in this case). For the scatter plot, it would be `scale_color_brewer(palette= "Dark2")`.

There are other packages that generate color palettes you can use in ggplot2. The viridis package offers functions similar to RColorBrewer's, but with different palettes; the ggthemes package includes palettes such as those that mimic color schemes used in Excel, Stata, Tableau, and the Economist. Look at the help pages for each of those packages to get more information about available palettes.

So far, these examples have been what are known as "discreet" variables: data that have separate and distinct categories, such as airports or states. There are a finite number of airports or states that can be in a data set. Color palettes can behave a bit differently when they're used with numeric, "continuous" variables such as temperatures or test scores.

We'll look at a few more examples of customizing colors in upcoming chapters, but I won't go into all the options in depth. Customizing ggplot2 visualizations could be a book in itself (and in fact, several such books have been published), and we've got a lot of other R capabilities to cover. If you're interested in doing more with color schemes in ggplot2, do check out the help files for various `scale_` functions and palette packages,

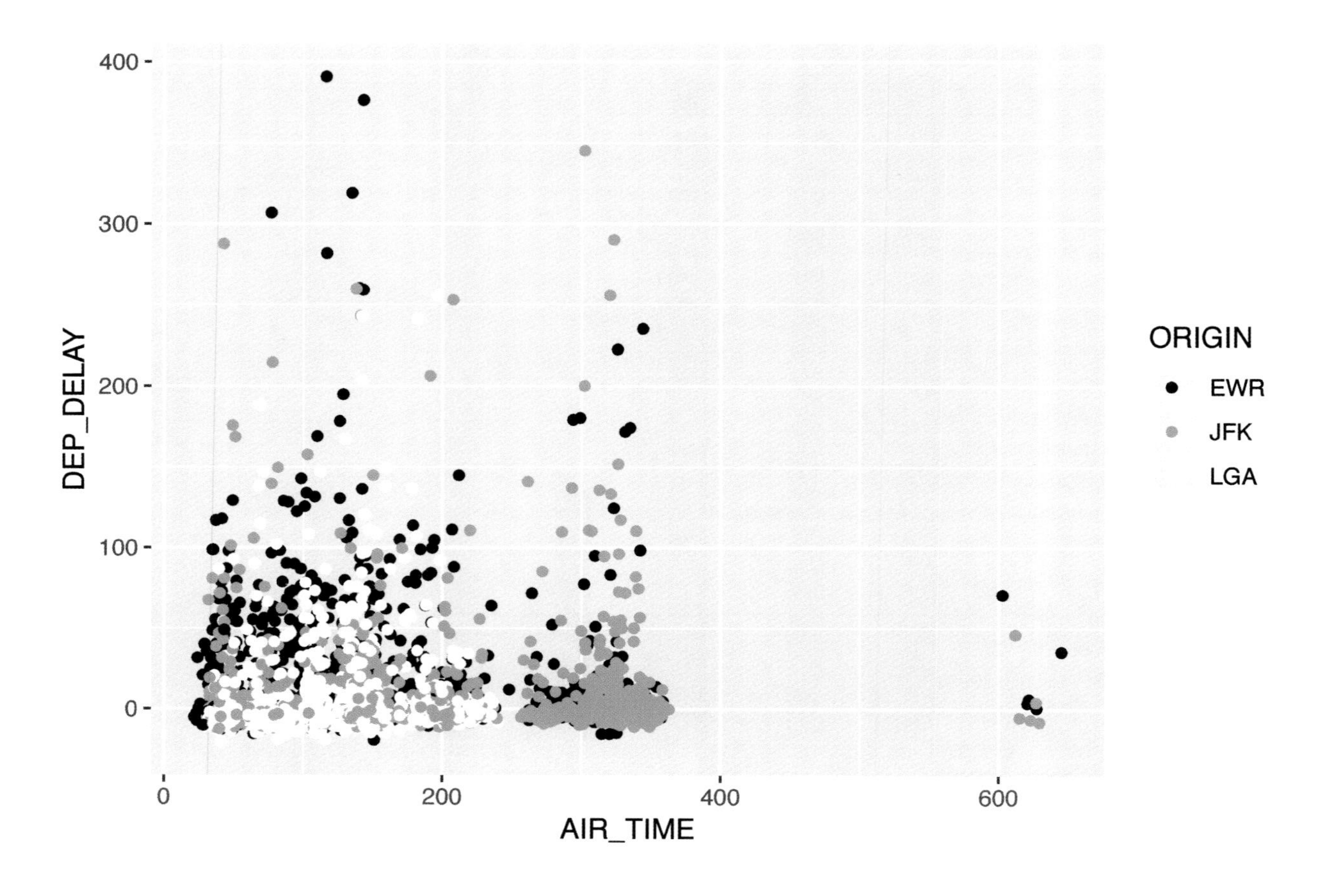

Figure 9.10: Scatter plot with customized colors.

as well as help pages in the ggplot2 online documentation under Scales at docs.ggplot2.org.

9.9 Other packages that extend ggplot2 functionality

As of ggplot2 version 2.0, there's a framework for creating "official extensions" to the library. Doctoral candidate Daniel Emaasit created a website to track available extensions; it's at http://www.ggplot2-exts.org/. Other packages create useful new themes. Still others are separate from the main project, but try to use similar syntax for other visualization types.

Among the most interesting:

- **ggiraph** – makes some ggplot2 visualizations interactive as HTML widgets.

This is a bit advanced to explain here, but if you'd like to see ggiraphs' capabilities, try running the following code on the lga dataframe from the previous chapter:

```
install.packages("ggiraph")
library(ggiraph)

lga$tooltip <- paste0(lga$CARRIER, " Flight ", lga$FL_NUM, " from ", lga$ORIGIN, " to ",
    lga$DEST, " on ", lga$FL_DATE)
lga$clickjs = paste0("alert(\"",lga$tooltip, "\")" )
```

```
myscatterplot <- ggplot(lga, aes(x = AIR_TIME, y = DEP_DELAY,
                  color = ORIGIN, tooltip = tooltip, onclick = clickjs)) +
    geom_point_interactive(position = "jitter") +
    scale_color_brewer(palette = "Dark2")

ggiraph(code = {print(myscatterplot)})
```

This may take awhile to run. If you mouse over any of the points on the resulting scatterplot in the Viewer (not Plots) tab, you'll see the underlying data. (Note: The paste0() function is just concatenating – that is, joining together – combinations of text strings and values. I'll explain more about that function in a future chapter. onclick and alert come from JavaScript.)

- **plotly** – This graphing library from Plotly also generates interactive graphics with a ggplotly() function.

This is a fairly easy way to create an interactive graphic from a ggplot2 dataviz, although you have to live with Plotly's style and behavior. Mouse over points to see underlying data; click and drag to zoom in; and use controls that you see when your cursor is within the graph, such as panning and zooming.

```
install.packages("plotly")
library(plotly)
myscatterplot <- ggplot(lga, aes(x = AIR_TIME, y = DEP_DELAY, color = ORIGIN)) +
  geom_point(position = "jitter") +
  scale_color_brewer(palette = "Dark2")
ggplotly(myscatterplot)
```

- **ggmap** – creates maps with base maps from Google Maps, OpenStreetMap, or other services and then uses ggplot2 syntax to map your data.
- **hrbrthemes** – adds ggplot2 themes that focus on typography, which can be particularly important for visualizations that will be published. It also includes gg_check() for spell-checking dataviz labels.
- **gganimate** – gives animation capabilities to plots over time. More info at https://github.com/dgrtwo/gganimate.
- **ggrepel** – helps with labeling items within a graph by ensuring text doesn't overlap, as it can do with use of geom_text().

You can see many more available ggplot2 extensions at the (unofficial) extensions Web site: www.ggplot2-exts.org.

9.10 Wrap-up

We learned how to use ggplot2 facets to create small, multiple graphs by category, as well as geofacets when creating geo-based multiple graphs. We also got an introduction to using color palettes with ggplot2, and a look at some packages that extend ggplot2 functionality.

9.11 Additional Resources

One example of a publication-quality graphic – with code – based on ggplot2 and geofacets: Bloomberg's visualization of enrollment in Affordable Care Act health insurance marketplaces (with some touch-ups in Adobe Illustrator). https://www.bloomberg.com/graphics/health-insurance-marketplaces-for-2018/

Len Kiefer, deputy chief economist at Freddie Mac, has done a lot of visualizing federal housing and economic data in R. He's got some interesting ideas for faceting, such as a post on visualizing housing-price trends at http://lenkiefer.com/2017/05/18/state-hpa. Kiefer's blog home page is lenkiefer.com.

Emil Hvitfeldt has been keeping a large list of available color palettes in R. That's on GitHub at https://github.com/EmilHvitfeldt/r-color-palettes.

geofacet package author Ryan Hafen wrote a blog post explaining the advantages of this dataviz type; see it at http://ryanhafen.com/blog/geofacet. The package Web site is at https://hafen.github.io/geofacet/.

Reminder that Additional Resources links are available in the booklinks.html file in this book's GitHub repo, so you don't have to type those lengthy URLs into your browser manually. You can also see them online at https://smach.github.io/R4JournalismBook/booklinks.html.

Next up: Create a map of median household incomes

9.12 Exercise 2 answer

Here's how you'd use summarize() to get a data frame of cancellation percentages:

```
library(dplyr)
library(ggplot2)
departing_cancellations <- ny %>%
  filter(ORIGIN %in% c("JFK", "LGA", "EWR"), FL_DATE %in%
  c("2012-10-27", "2012-10-28", "2012-10-29", "2012-10-30",
  "2012-10-31","2012-11-01", "2012-11-02", "2012-11-03")) %>%
  group_by(ORIGIN, FL_DATE) %>%
  summarize(
    Total = n(),
    NumCancelled = sum(CANCELLED),
    PctCancelled = round((NumCancelled / Total) * 100, 1)

  )
```

Bonus tip: I wrote out the specific dates around Sandy in that code above, which I've done in earlier chapters. That ignores a fundamental goal of good programming: Don't Repeat Yourself (DRY). At some point in this project, it would make sense to define a vector of sandy_dates, such as `sandy_dates <- c("2012-10-27", "2012-10-28", "2012-10-29", "2012-10-30", "2012-10-31","2012-11-01", "2012-11-02", "2012-11-03"))` and re-use those. That would make the filter line above `filter(ORIGIN %in% c("JFK", "LGA", "EWR"), FL_DATE %in% sandy_dates)`.

(I could also do the same with the three airport codes, storing the NY airports in a vector such as `ny_airports <- c("JFK", "LGA", "EWR"), FL_DATE %in% sandy_dates)`.)

Chapter 10

Write your own R functions

At some point in your R journey, you're going to want a function that's not available in an existing package in the exact format you need. Which means it's time to write your own.

This may sound like an advanced skill, but it isn't. In some cases, you may simply be tweaking an existing function to fit a current project. But even if you're writing a function from scratch, I think you'll find the benefits-to-difficulty ratio to be very high.

So let's get started.

10.1 What we'll cover

- Writing simple functions
- An introduction to date objects
- Useful base R functions `cut()` and `seq()`
- A shortcut to create character vectors, Hmisc::Cs()
- If-then conditions in R
- A peek at automated testing

10.2 Packages needed in this chapter

```
pacman::p_load(dplyr, lubridate, testthat)
```

10.3 Function basics

Functions don't have to be complicated. Basic function syntax is as easy as

```
myfunction <- function(x){
  my_result <- some_code_i_run(x)
}
```

Let's go over each piece of this code. `myfunction` is the name I gave the function – I could have called it pretty much anything, as long as I followed the rules around R variable names. In general, you'll be fairly safe naming functions if you follow some beginner's basics: Start with a letter followed by letters, numbers, dots, or underscores; and don't use a name that already does something else in R. (`?make.names` shows more complete rules).

`function()` says that the myfunction variable is, well, going to be a function and not simply storing a vector or data frame. That part *has to* say `function`, and `function()` *has to* have parentheses after it.

The parentheses contain whatever *arguments* the function uses. Arguments are variables that are used when functions are run. If you run `sqrt(9)` to get the square root of 9, `9` is the argument of the sqrt() function.

A function can have zero arguments, if you just want the function to do the same thing over and over again. For example, here's a function that prints "Today is" and the current date:

```
printToday <- function(){
  my_text <- paste0("Today is ", Sys.Date())
  print(my_text)
}
```

This function will always do the same thing on the same date: Print "Today is" and the system date.

As I mentioned last chapter, the paste0() function just concatenates – joins together – items into a single string. I'll cover this in a bit more detail next chapter.

To try out the printToday function, run the code above to *create* the function, and then run `printToday()` to *run* the function. The parentheses are needed, even without any argument. If you run `printToday` *without* the parentheses, R will show you the code that created the function instead of actually running the function. Running a function without parentheses is handy if you want to see how someone else's function works under the hood.

```
printToday()
```

```
## [1] "Today is 2018-06-20"
```

```
printToday
```

```
## function(){
##   my_text <- paste0("Today is ", Sys.Date())
##   print(my_text)
## }
```

You might have noticed that `Sys.Date()` is also a function that takes no arguments.

In this next example, the printFuture() function prints today's date plus x number of days, so it takes one argument: the number of days to add to today's date.

```
printFuture <- function(x){
  my_text <- paste0("Today plus ",  x, " days is ", Sys.Date() + x)
  print(my_text)
}
```

```
printFuture(7)
```

```
## [1] "Today plus 7 days is 2018-06-27"
```

You may know that you'll often want a function's argument to have one specific value – for example, using the printFuture() function to return the value for tomorrow (and so x would equal 1) – but you also want the option to change that value. In that case, you can set a default value for the function's argument.

The syntax for setting a default is:

```
myfunction <- function(x = my_default_value){
  my_result <- some_code_i_run(x)
}
```

So, for printFuture, the first line of the function definition would be `printFuture <- function(x = 1){` if you want the default to be tomorrow's date:

```
printFuture <- function(x = 1){
  my_text <- paste0("Today plus ",  x, " days is ", Sys.Date() + x)
  print(my_text)
}
```

A function can have more than one variable, such as `function(x, y)`. (Advanced tip: `...` in the function definition means that you can have as many arguments as you want, without having to specify the exact amount. For example, sum() adds up as many numbers as you give it.)

That should be enough to get you started writing functions in R, although there are a lot of more advanced concepts to make your functions more efficient and well structured.

If you've written functions in other programming languages, I want to mention a couple of things that might be different from what you're used to. While you do have to explicitly state the arguments you want in an R function, you *don't* have to specify what *type* they are (character strings, numeric, etc.). That's not the case in all programming languages. And, a function will return the value of its last line. If you want some other value, use `return(myvalue)`; but return() isn't always mandatory, as it is in some other languages. If you want the function to print out its result when it's not being assigned to a variable, though, use `return(myvalue)`.

There are some other important issues around functions, like whether or not variables created inside a function are available outside that function for use in other code. However, that's going beyond the scope of this book. (Yes, that pun was intended: Scope is a programming concept that defines where variables can be accessed.) For now, it's best to assume that anything going on inside a function *is only available within the function brackets*, and only the final line's value can emerge out of the function for use elsewhere.

10.3.1 Custom functions with dates

Here's a scenario where a function might come in handy: You want to write a story each month about local unemployment rates, or crime rates, or a local stock index you created. Imagine that you can pull data automatically with a script (not being written here); but to fully automate the process, you want a custom function that will generate the date for "the beginning of last month".

We'll learn more about dates in R in an upcoming chapter. For now, I'll give you these basics:

- `Sys.Date()` gives you today's date, as seen earlier in this chapter.
- `months()` is a base R function that allows date calculations by months. You can add 2 months to a date with `+ months(2)` or subtract 1 month from a date with `- months(1)`.
- `cut()` is another base R function. It separates a range of values into intervals. cut() has some special features when used with R date objects: It can return the beginning of a "month" as well as a "quarter", "year", and several other intervals.

If x is an R date object, `cut(x, "year")` will return Jan. 1 of that year; and cut(x, "month") will return the first of that date's month.

That means in order to get the first of *last* month, we could 1) Determine today's date, 2) Calculate today's date minus a month, and 3) get the first of the month for today's date minus one month. Here's my function:

```
get_first_of_last_month <- function(mydate = Sys.Date()){
  # Gets the input date, which defaults to today, and subtracts 1 month
  mydate <- mydate - months(1)
  mydate <- cut(mydate, "month") # Finds the 1st of the month for that date
  return(mydate)
}
```

I added `return(mydate)` so that the function will print its result in the console if I run the code without storing the result in a variable.

If you run that code to create the function and then run the function with `get_first_of_last_month()` you should see the first of last month returned in your console.

There's one more point here, though, which is important when your function is returning a value: Is that value the class – in other words, the *type* – that you expect?

If you ran the code, you probably saw that the return value is actually a *factor* and not either a date object or a character string. That's because cut() is designed to create levels from your data. Run cut() on several dates and it will be easier to understand:

```
mydates <- as.Date(c("2016-02-27", "2016-02-28", "2016-02-29", "2016-03-01",
                     "2016-03-02", "2016-03-01" ))
cut(mydates, "month")
```

```
## [1] 2016-02-01 2016-02-01 2016-02-01 2016-03-01 2016-03-01 2016-03-01
## Levels: 2016-02-01 2016-03-01
```

The result gives you the beginning of the month for each date, and creates factors (categories) for those levels. This makes sense because the whole point of cut() is to create categories from data. However, that's probably not what you want when working with dates. You'll probably want either character strings or R Date objects.

You can make the function return a date by changing the last line of the function to `return(as.Date(mydate))` or a character string with `return(as.character(mydate))`.

Now you've got a function to get "the first of last month".

For more information on using cut *specifically with dates,* run `?cut.Date` for that help file.

10.4 seq()

There's an easier way to get a series of dates than writing them out the way I did, with `as.Date (c("2016-02-27", "2016-02-28", "2016-02-29", "2016-03-01", "2016-03-02", "2016-03-01" ))`. R's `seq()` function generates a series of values using the syntax `seq(from, to, by)`. `from` is the start, `to` is the end, and `by` is the interval desired (1 if you want every value, 2 for every other value, etc.). An alternate syntax is `seq(from, length.out, by)` where length.out is the number of items you want in the series.

Like the cut() function, seq() has a special version for date objects, seq.Date(). `seq.Date(as.Date ("2018-02-27"), length.out = 6, by = "day")` would generate a series of six date objects starting on February 27, 2018: See `?seq.Date` for more information, including the various `by` intervals supported, such as "day", "week", "month", and "quarter".

Exercise 3: How would you write a function to get the *last day* of last month?

10.5 If-then-else

Now suppose you need to get the date for "Monday of last week". If you know you run this function every Monday, it would be easy enough to just subtract 7 days from "today": `Sys.Date() - 7`. But what about the weeks with Monday holidays, when you need to run your script on Tuesday to find "Monday of last week"?

A labor-saving solution would be setting up the script on a server that automatically triggers your script every Monday, whether or not you're working. (Or, if you're like many journalists I know, the other solution would be to run the script every Monday, even on your days off.)

But for the sake of this exercise, assume that you actually do take Monday holidays off, and this project can't be fully automated on a server, so your function needs to find "Monday of last week" on other days.

Let's first look at the simplest case: You're *100% sure* that you'll run the script either on Monday or Tuesday for the prior week. If you run the function on a Monday, you want the result to be "today minus 7". The only other possibility is Tuesday, when the result should be "today minus 8".

In its simplest form, R's `if` statement looks like `if(condition) result`. And by simplest, I mean the result can be written in a single line of code.

I'll store today's date in a variable called today, and then run a simple `if()` statement. And, I'll use base R's weekdays() function to determine the current day of the week.

(Aside: Base R's weekdays() function is similar to the Excel formula `=TEXT(A1, "dddd")`. Which would you say is more intuitive to get day-of-week information as an English word?).

```
today <- Sys.Date()
if(weekdays(today) == "Monday") {
  lastMonday <- today - 7
}
```

The code after the first line says: If the day of week for the today variable equals Monday, then set the value of the variable lastMonday to today - 7 days.

Did you remember the == operation from Chapter 5? If you're *testing whether one object equals another* you need == and not =. A single equals sign *sets a value*; two equals signs perform a logical test.

If you run that code, a value will be set for lastMonday *only* if today happens to be Monday. If today *isn't* Monday, nothing will happen and no value will be set.

To add a default for situations when the if statement is false, you need an `else` statement. The general format should be

```
if(condition){
  some code
} else {
  some other code
}
```

The placement of else is important: `else` *must be on the same line as that first closing bracket.*

So now we can say that if today is Monday, set lastMonday to today - 7; otherwise, set lastMonday to today -8:

```
today <- Sys.Date()
if(weekdays(today) == "Monday") {
  lastMonday <- today - 7
} else {
  lastMonday <- today - 8
}
```

If you want to see whether this code works properly, set `today <- "Monday"` manually and then `today <- "Tuesday"` manually.

You probably see the glaring problem here, though. This kind of if-else statement works great when you are absolutely, positively, 100% sure you have a binary condition. "If a coin flip result is heads, do something; otherwise, do something else" is pretty dependable code, since you're unlikely to end up with a third choice of a coin balanced on its side.

In this case, though, you might be incredibly certain that you'll run this code on a Monday or Tuesday, but life could have other plans – you were out sick Monday and Tuesday, or took a few days off. And while this is a fairly obvious problem, sometimes all possible options aren't as clear while you're writing your function. In general, when writing if statements, it's important to consider whether you've accounted for every possible situation.

A quick-and-dirty way to account for other days is to simply classify anything that's not "Monday" or "Tuesday" with a warning message by using two if-else conditions and then a final else with an R warning() statement:

```
today <- Sys.Date()
if(weekdays(today) == "Monday") {
  lastMonday <- today - 7
} else if(weekdays(today) == "Tuesday") {
  lastMonday <- today - 8
} else {
  lastMonday <- warning("You are running this script on a day that isn't
                Monday or Tuesday. Results will not be accurate.
                Please calculate the date for last Monday manually.")
}
```

```
## Warning: You are running this script on a day that isn't
##                Monday or Tuesday. Results will not be accurate.
##                Please calculate the date for last Monday manually.
```

Now that you've got the basic code, you can put that code into a function by wrapping it in `get_last_Monday <- function() {}` (and setting a default value for "today"):

```
get_last_Monday <- function(the_day = Sys.Date()){
  the_day <- Sys.Date()
  if(weekdays(the_day) == "Monday") {
    lastMonday <- the_day - 7
  } else if(weekdays(the_day) == "Tuesday") {
    lastMonday <- the_day - 8
} else {
    lastMonday <- warning("You are running this script on a day that wasn't
                      Monday or Tuesday.")
  }
}
```

That will work on other days, in as much as it won't give you a wrong date. Ideally, though, you'd like your function to handle all possibilities. In this case, that would be *all* days of the week, not just Monday or Tuesday.

You could handle this with a lot of `else if` statements or a function in R called switch(), but I've become a fan of dplyr's case_when() function as being an elegant, compact, and intuitive way to do this.

case_when() takes the format:

```
case_when(
  condition1 ~ result1,
  condition2 ~ result2,
  condition3 ~ result3,
  ...
)
```

Note that the operator between the condition and the result is a ~ tilde sign, not equals.

So, in this, um, case, assuming we want the result to be "the Monday before the most recent Monday":

```
get_last_Monday <- function(the_day = Sys.Date()){
  today <- weekdays(the_day)
  my_result <- dplyr::case_when(
    today == "Monday" ~ the_day - 7,
    today == "Tuesday" ~ the_day - 8,
    today == "Wednesday" ~ the_day - 9,
    today == "Thursday" ~ the_day - 10,
    today == "Friday" ~ the_day - 11,
    today == "Saturday" ~ the_day - 12,
    today == "Sunday" ~ the_day - 13
  )
  return(as.character(my_result))
}
```

Each logical test for today is on the left side of the ~ tilde and its result is on the right.

As you look at that code, though, you might notice it's an awful lot of repetition of a pattern: -7, -8, -9, and so on. I implemented the code using case_when() to show you how the function works. But a good programmer would try to simplify that. Is there some way to express the results as a numerical pattern?

It would help to get a *number* for the day of the week instead of the name. The lubridate package's wday() function does this. If you install and load lubridate the usual ways (install.packages() and library() or pacman::p_load()), you can use wday(Sys.Date()) to get 1 back if today is Sunday, 2 back for Monday, and so on.

I'll create a data frame so you can better see the pattern – and also see one of my favorite shortcuts, the Hmisc package's Cs() function. Normally, to create a character vector, each of the words/character strings needs to be inside quotation marks:`c("Monday", "Tuesday", "Wednesday")` and so on. Cs() turns (Monday, Tuesday, Wednesday) into c("Monday", "Tuesday", "Wednesday"). So, with Cs:

```
DaysofWeek <- Hmisc::Cs(Monday, Tuesday, Wednesday, Thursday, Friday, Saturday, Sunday)
DaysofWeek_Numeric <- c(2,3,4,5,6,7,1)
NumberToSubtract <- c(7,8,9,10,11,12,13)
thepattern <- data.frame(DaysofWeek, DaysofWeek_Numeric, NumberToSubtract,
stringsAsFactors = FALSE)
print(thepattern)
```

```
##   DaysofWeek DaysofWeek_Numeric NumberToSubtract
## 1     Monday                  2                7
## 2    Tuesday                  3                8
## 3  Wednesday                  4                9
## 4   Thursday                  5               10
## 5     Friday                  6               11
## 6   Saturday                  7               12
## 7     Sunday                  1               13
```

Exercise 4: Create a function that uses the numeric weekday values to return the date for "Monday of last week," accounting for possibilities for all seven weekdays – *without* explicitly writing out each case.

10.6 if statements for vectors

An if statement works fine for checking a condition for *a single value*, but you'll run into problems if you want to use it for checking *multiple* values. See what happens when I try to create a function to check if a number is odd or even. The code below creates a function that accepts one value (a number) and calculates whether the remainder is 0 or 1 if the number is divided by two. (%% is R's modulo operator to get the remainder after dividing one number by another.)

```
OddOrEven <- function(thenumber){
  if(thenumber %% 2 == 0){
    result <- "Even"
  } else if(thenumber %% 2 == 1){
    result <- "Odd"
  } else {
    result <- "Neither even nor odd"
  }
  return(result)
}
```

Checking one number works fine, such as `OddOrEven(376)`. But if you have a vector of numbers, such as `mynumbers <- c(376, 245, 12)`, trying `OddOrEven(mynumbers)` will throw a warning that `the condition has length > 1 and only the first element will be used`.

if() will only work on one set of conditions, such as whether 4 is odd or even. To run if or if-else statements on more than one value, you need R's `ifelse()` function. The format is `ifelse(condition, value if condition is true, value if condition is false)`. I could create a new function, IsItOdd:

```
IsItOdd <- function(thenumber){
  result <- ifelse(thenumber %% 2 == 1, "Odd", "Not odd")
  return(result)
}
```

That ifelse statement says "If thenumber divided by 2 leaves a remainder of 1, thenumber is Odd; otherwise thenumber is Not odd." This function will now work on a vector, such as `IsItOdd(mynumbers)`. In R programming-speak, it's been "vectorized", so you don't need a function like lapply() or purrr's map() to run it on multiple values at once.

ifelse() is binary – there's one value for when the test is true and another value for when the test is false. ifelse() gets a bit harder to follow if you want a function with more than two options. That's why I tend to use case_when(), which handles multiple options and also has a catch-all `TRUE` for everything else:

```
IsItOddOrEven <- function(thenumber){
  result <- case_when(
    thenumber %% 2 == 1 ~ "Odd",
    thenumber %% 2 == 0 ~ "Even",
    TRUE ~ "Not an integer"
  )
  return(result)
}
```

This will now work for `IsItOddOrEven(c(5,8,3.14))`, although it will still throw an unpleasant error if someone uses it for a character string. Exercise 5: To fix that, you could use an if-else to first check that thenumber is either an integer or numeric. If you'd like to give that a try, the answer is at the end of this chapter.

10.7 A taste of testing

Change one line of code, and you can screw up something else without realizing it. That's one of many reasons why scripting and automating *tests of your code* is something worth learning.

test_that is "an R package to make testing fun," according to creator Hadley Wickham. I can't guarantee that you'll have fun adding tests to your code, but at the very least, it should makes the process less annoying.

If you're interested in knowing a bit about testing now, read on. Otherwise, feel free to skip to the next section and come back here when you think you want to add testing to your coding arsenal.

10.7.1 An incredibly brief look at R testing

When using testthat, you start with a small check of some condition.

For example, if I ran the get_last_Monday() function on February 14, 2018, I should get back "2018-02-05" as a result – that's Monday in the prior week. And, if I wrote my function so last Monday's date is supposed to be returned as a character string and not a date object, I'd expect the result to be type "character."

Each of those statements could be considered a testthat "expectation."

A *test* is one or more of those expectations. An example should (hopefully) make this clearer.

The code below uses testthat's test_that() function to create a test called "get_last_Monday_results." This test checks for three expections: whether my function returns "2018-02-05" if I run it on Feb. 14, 2018 and Feb. 15, 2018, and whether the result is a character string.

```
library(testthat, quietly = TRUE)
test_that("get_last_Monday results", {
  expect_equal(get_last_Monday(as.Date("2018-02-14")), "2018-02-05")
    expect_equal(get_last_Monday(as.Date("2018-02-15")), "2018-02-05")
  expect_type(get_last_Monday(), "character")
})
```

If these expectations are met, *nothing happens*. If the expectations *aren't* met, the test will return a failure message and explain what failed, such as `* get_last_Monday(as.Date("2018-02-14")) not equal to "2018-02-05"`. You can also store the results of test_that() in a variable, which will be TRUE if all expectations are met and FALSE if at least one fails.

This is a trivial example. But as your projects become larger and more complicated, being able to check that everything works as you expect after changing some code is enormously helpful. If the tests pass, you can feel more confident about your scripts without having to re-do some testing manually. And if a test fails, it can help you narrow in on where the problem might be.

Finally, testing helps you think more methodically about your code. "What should each small block of code do, exactly?" is a great question to keep in mind as you're working to automate your analysis.

If you want to find out more about testing, Testing R Code by Richard Cotton (CRC Press) is a great resource. In addition, Hadley Wickham has a chapter on tests in his R Packages book, available free online at http://r-pkgs.had.co.nz/tests.html although, as the title implies, this is focused on testing for developing R packages and not stand-alone functions.

10.8 Next steps for your functions

Sometimes, a custom function that's useful for one specific project will be useful for others. You don't want to be cutting and pasting multiple copies of a function in different project files. What if you need to make a change in the function? You don't want to find copies of the same code scattered all over your system.

There are a few ways to handle this.

The easiest is to simply have a directory of useful R functions on your computer. Then, you can have one or more files of your favorite functions, and just use `source("path/to/myfunctionsfile.R")` code in your scripts to load everything in that myfunctionsfile.R file into your working sessions. However, there are a couple of problems with this approach. What if, like me, you do R work on more than one computer? And what if you'd like others to be able to run your code and check your work?

If there's nothing sensitive in your functions such as API keys, passwords, or company-confidential data, you can store your files in the cloud. A GitHub "gist" is designed for easy code access. You can create a free GitHub account and then click the Gist button. Once you create a gist, you'll see its URL. To access the *code-only URL,* click the Raw button on the gist. If you made it public, you or anyone else can access it in R at that Raw URL with a line of code such as `source("https://gist.githubusercontent.com/yourusername/gistid1/raw/gistid2/gistname.R")`.

Once you're an experienced R user, though, one of the best ways to handle functions that you're likely to use a lot is to create your own R packages. That's beyond the scope of this book. However, if and when you're ready, Etsy data analyst Hillary Parker has a wonderful and easy-to-follow blog post at https://hilaryparker.com/2014/04/29/writing-an-r-package-from-scratch/. "This tutorial is not about making a beautiful, perfect R package," she explains. "This tutorial is about creating a bare-minimum R package so that you don't have to keep thinking to yourself, 'I really should just make an R package with these functions so I don't have to keep copy/pasting them like a g–damn luddite.' "

How can you argue with that?

Next up: Mapping in R.

10.9 More Resources

Hadley Wickham and his sister Charlotte Wickham have an interactive video course on DataCamp about writing functions: https://www.datacamp.com/courses/writing-functions-in-r. The first part of the course is free, and covers a fair amount including scoping (understanding environments within and outside of a function). Access to the rest of the class requires a $29/month paid account.

Hadley Wickham also has a free online book about how to create R packages, at http://r-pkgs.had.co.nz/.

For a quick look at code testing in action, see my screencast R tip: Test your code with testthat at https://www.infoworld.com/video/87735/r-tip-test-your-code-with-testthat. I also recorded a screencast showing how to use the case_when() function, at https://www.infoworld.com/video/87435/r-tip-learn-dplyr-s-case-when-function.

Finally, there are more compact ways to calculate "most recent Monday" than the examples I used in this chapter. This StackOverflow question has several options: http://stackoverflow.com/questions/32763491/find-most-recent-monday-for-a-dataframe.

10.10 Exercise 3 Answer

One way to create a function that returns the last day of last month is to find the first of "this month" and then subtract 1 day. In base R, subtracting 1 day from a date object can be shortened to simply subtracting 1:

```
get_end_of_last_month <- function(the_day)
{
    firstmonth <- as.Date(cut(the_day, "month"))
    endlastmonth <- firstmonth - 1
    return(endlastmonth)
}
```

10.11 Exercise 4 Answer

The pattern looks quite straightforward for every day except Sunday: NumberToSubtract is 5 less than DaysofWeek_Numeric. Sunday is the lone exception. Looks like a perfect use of an if -else condition:

```
get_last_Monday <- function(the_date = Sys.Date()){
  today_dayofweek_numeric <- lubridate::wday(the_date)
  today_dayofweek <- weekdays(the_date)


  if(today_dayofweek != "Sunday"){
    NumberToSubtract <- today_dayofweek_numeric + 5
  } else {
    NumberToSubtract <- 13
  }

lastMonday <- the_date - NumberToSubtract
return(as.character(lastMonday))
}
```

10.12 Exercise 5

Checking whether the function's argument is numeric before running case_when():

```
IsItOddOrEven <- function(thenumber){
  if(is.numeric(thenumber) | is.integer(thenumber)){
  result <- case_when(
    thenumber %% 2 == 1 ~ "Odd",
    thenumber %% 2 == 0 ~ "Even",
    TRUE ~ "Not an integer"
  )
  } else {
    result <- "This function needs arguments that are integers or numeric."
  }
  return(result)
}
```

Chapter 11

Maps in R

R may never replace full-featured mapping software like Esri's ArcGIS or open-source QGIS, which were specifically designed for sophisticated geospatial analysis. However, R does have some pretty robust capabilities for working with spatial data. And, recent years have seen some new mapping options that are both powerful and easier to use than prior iterations.

Why map in R? Honestly, I find Google Fusion Tables easier to use for simple mapping displays. However, that's not an open platform, so you're at the mercy of Google's desire to keep supporting it. In addition, Fusion Tables means opening a separate application if you're already doing other data import and analysis in R. (And, Google mapping within Fusion Tables has its limits). I like R when I want a single, reproducible workflow; or when I'm looking for more customization than Fusion Tables or Google Maps offers out of the box. There's a lot of customization you can add to Google maps if you write some JavaScript code, but at that point it's not necessarily any easier than R.

11.1 Map projects in this chapter

California median income by county bank branches in Boston demographic information + bank branches in Boston mapping new boundary districts

- California median income by county
- Bank branches in Boston
- Demographic information + bank branches in Boston
- Mapping new boundary districts

11.1.1 Which packages for mapping in R?

```
pacman::p_load(leaflet, glue, dplyr, sf, tmap, tmaptools, tidycensus, ggmap, htmltools,
htmlwidgets)
pacman::p_load_gh(c("walkerke/tigris", "bhaskarvk/leaflet.extras"))
```

Note: If you have trouble installing glue with pacman::p_load(glue), try `install.packages("glue")`. And make sure rio is installed.

You'll also need a U.S. Census *API key* if you don't have one. Request a free key at http://api.census.gov/data/key_signup.html. An API, or Application Programming Interface, is a service designed for developers to get data by writing some code. Some APIs are open; others require a key (often to make sure that a single user isn't hogging too much bandwidth). You can find out more about getting and using your key by running `?census_api_key`.

Like most things in R, your choice of package may depend as much on personal syntax preference as it does on what you're trying to accomplish, as well as where you fall on the ease-of-use vs. need-for-power spectrum.

I tend to use the tmap package for creating static, color-coded theme maps, but you won't go wrong if you go with other options like choroplethr or ggmap and ggplot2.

For *interactive Web choropleth (color-coded) maps,* I find tmap a good choice for ease of learning, while the leaflet package has more capabilities for customization. If you're serious about creating interactive choropleth maps for Web publication, learning leaflet, an R interface to the popular Leaflet JavaScript library, would be a worthwhile time investment.

To create an interactive map with *point markers*, leaflet is considerably easier to use than when making choropleths.

And, if you want the capabilities of a full-fledged Geographic Information System, there's an R package to interface with QGIS as well as a "bridge" between ArcGIS and R.

In short: You can do quite a lot with spatial data in R. There's a lot to sample!

11.2 Skills we'll cover

- Importing geographic shapefiles into R
- Finding ready-to-use U.S. Census shapefiles that include data
- Generating static and interactive choropleth maps
- Joining geospatial files with other data
- Geocoding addresses with ggmap
- Learning R's paste() and paste0() functions
- Looking at glue, a paste0 alternative
- Creating an interactive map of geocoded locations
- Combining points and polygons on a map
- Adding address search to a map

11.3 Importing shape files into R

There are several popular formats for defining geographic boundary lines. Shapefiles, created by Esri, are now used in many other applications besides Esri's own software. KML files are the format of choice at Google. And GeoJSON is another standard used by some software and APIs. For projects in this chapter, we'll use shapefiles.

For many mapping analyses, you need 1) a file defining a geographic area such as towns, counties, or states; 2) a file with data about those units, such as which towns voted for what candidates; and 3) a way to merge the two, and then display the results.

For geographic boundary files of U.S. areas such as states, counties, Congressional districts, state legislative districts, or ZIP codes, the U.S. Census Bureau has shapefiles available for download. Look up Census TIGER shapefiles on Google, and you should be directed to the Census Bureau's available files. (At the time I wrote this, the files were at https://www.census.gov/geo/maps-data/data/tiger-line.html).

However, you can also import Census Bureau shapefiles directly into R with the tigris package's GitHub version. The tigris package is on CRAN, but if you install the CRAN version, that version may be considerably older than what package author Kyle Walker has on GitHub.

tigris has some handy functions making it easy to find geographic files. For this chapter's first project, I'd like to map housing prices by county in California. So, I first need a shape file with California counties. Downloading a shapefile of California counties and saving that to a variable called ca_counties is as easy as

```
ca_counties <- counties("CA")
```

You can do a quick check to see if you've gotten what you think you did by plotting ca_counties with base R's generic plot() function (see Figure 11.1):

```
plot(ca_counties)
```

Figure 11.1: Simple plot of the California counties shapefile.

There are a lot of helper functions in tigris for downloading shapefiles. You can see the full list on the package's GitHub repository at https://github.com/walkerke/tigris. If, for example, you wanted a shape file of California school districts:

```
ca_schools <- school_districts("CA")
```

tigris has files from 2011 on, but not all files are updated to current years. So, be careful if you're working with something like legislative districts that may have changed since the most-recent available year in tigris.

You can specify the year you want in tigris with the `year` argument, such as `ca_counties <- counties("CA", year = 2015)`. You'll get an error message if you try to download a file that's not available.

By default, these files come into R as a *SpatialPolygonsDataFrame,* which has been a go-to format for R maps for some time. However, more recently, the *simple features* mapping standard is also available to use in R. This is excellent news. I find SpatialPolygonsDataFrames to be more complicated to work with than conventional data frames, while simple features *are* conventional data frames that have one special column of geography.

tigris can create a simple-features sf object by adding the `class="sf"` argument when downloading files:

```
ca_counties_sf <- counties("CA", class = "sf")
```

You can check out the difference between a SpatialPolygonsDataFrame and a simple-feature data frame by running either str() or dplyr::glimpse() on both objects. (glimpse can be a good choice for larger, more complex objects). ca_counties_sf looks like a typical data frame; ca_counties will look a lot more complex.

11.3.1 Importing shapefiles from anywhere

Of course, a lot of mapping projects need shapefiles that aren't available from the U.S. Census Bureau – such as analysis involving any non-U.S. geography, or mapping U.S. city neighborhoods or voting districts. Some other good sources of shapefiles: Global Administrative Areas http://www.gadm.org/country (which actually already has R SpatialPolygonsDataFrames as well as shapefiles), Eurostat https://www.gadm.org, Statistics Canada https://www12.statcan.gc.ca/census-recensement/2011/geo/bound-limit/bound-limit-eng.cfm, ArcGIS Online at arcgis.com (ArcGIS Public free account needed to open and download), Zillow for U.S. city neighborhood boundary maps at https://www.zillow.com/howto/api/neighborhood-boundaries.htm (note the license restrictions requiring credit to Zillow), and numerous national, state, and local agencies. If you can't find what you want online, check with the planning or engineering departments in the area involved (that's how I got a shapefile of new voting districts in my town).

You can import a non-Census-Bureau shapefile into R using several different packages and functions. The tmaptools package's read_shape() function is pretty easy to use.

Some mapping software lets you import shapefiles while they're still zipped; but for tmaptools' read_shape(), you'll need to unzip them first if they came zipped. For unzipped shapefiles, the format for import is `mygeo <- read_shape("path/to/shapefile/myshapefile.shp")` to read a shapefile into a variable. Don't forget the .shp at the end of the file name.

Next, it's helpful to map what you've imported. tmap has a quick thematic mapping function qtm() that's similar to ggplot2's qplot(). And, like ggplot2, tmap also has a more full-featured visualization function that I'll cover later in this chapter.

To generate a basic map of a shapefile imported into a variable called mygeo, the syntax would be `qtm(mygeo)`.

qtm() can generate a basic map of the ca_counties_sf object (see Figure 11.2) with:

```
qtm(ca_counties_sf)
```

Note: When analyzing spatial information in R, the datum *and* coordinate reference system *(CRS) of your data files are very important. To simplify, datum accounts for the fact that the earth isn't exactly round, and locates points based partly on a specific model of the earth's actual shape. The CRS solves a different problem: Your map is two-dimensional but the earth is a 3-D globe. Delving into these concepts is far beyond what this entry-level mapping introduction can cover. But you should know that for any analysis where exact borders and point placement matter, you need to use consistent and correct values for both. If you don't understand how to do this, ask for help from an expert.*

If you want to learn more about those concepts, or already are familiar with them but want to know how to deal with them in R, I've included some resources at the end of this chapter. In some of these exercises, I've assumed that when creating color-coded maps of counties or towns, it's probably not a problem if boundaries end up a couple of hundred meters off. For GIS analysis where precise locations are important, though, such as calculating how many people live within a mile of an aging dam, it's important that all data are using the same datums and projections.

11.4 Import data for mapping

The next step for making a choropleth map is having some numerical data about geographic units, in this case information about California counties. There are a lot of interesting options – median household income, median housing prices, unemployment rates.... I'll start with median income, which is also available from the Census Bureau.

There's a package, tidycensus, that makes all this incredibly easy, which I'll cover in the next section. But please don't skip this part. At least skim it! Because the day will likely come when you'll want to map data that's not conveniently packaged by the Census Bureau, and this skill will be important.

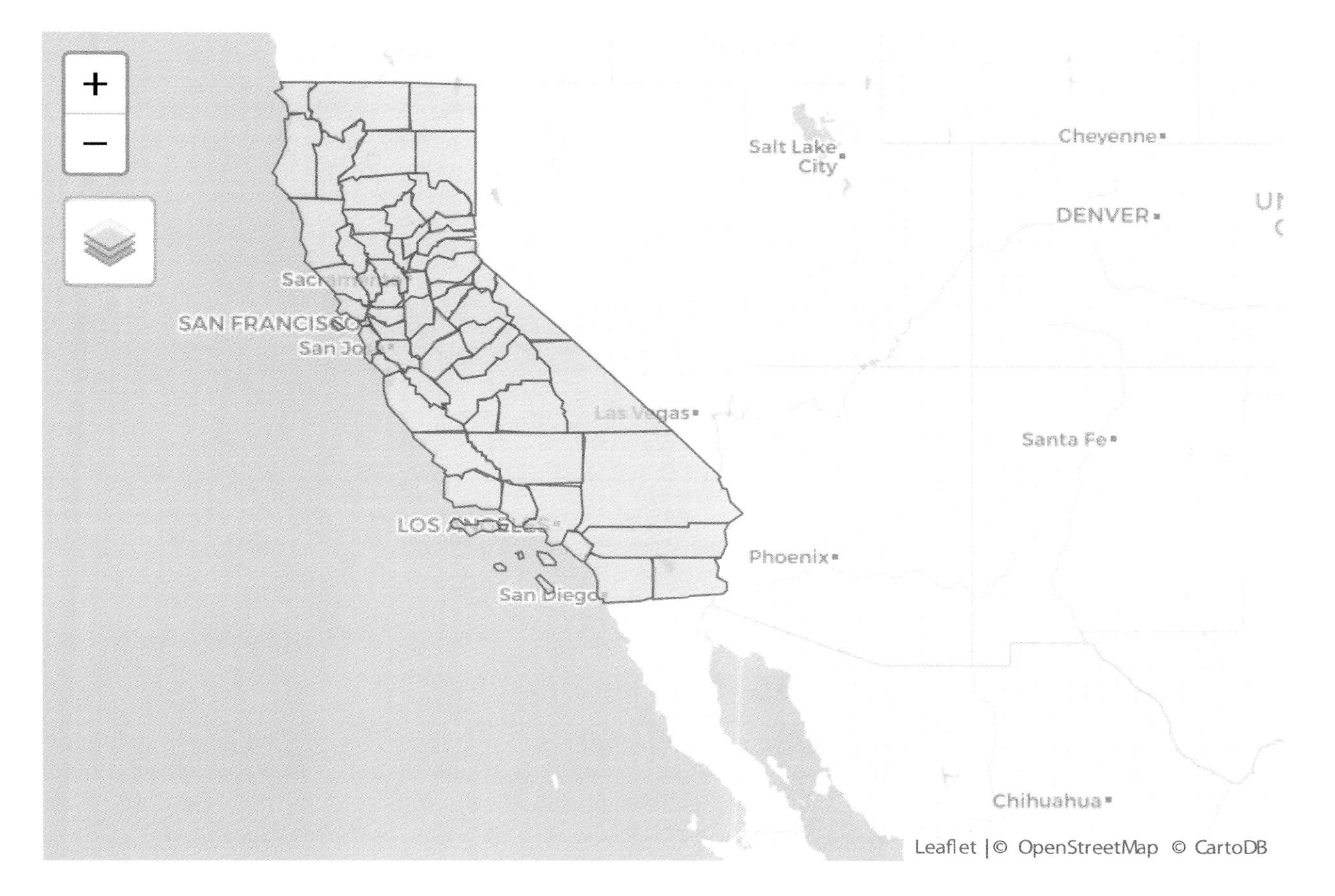

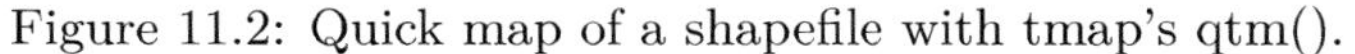

Figure 11.2: Quick map of a shapefile with tmap's qtm().

One easy-to-use sources for US Census information is the Census Reporter site at censusreporter.org. Head over there and search for median household income in the Explore box; the first table to appear in the dropdown list should be Median Household Income. See Figure 11.3.

You can select that table from the dropdown list by clicking on it. Next, type in `California` where it says "Start typing to pick a place," choose counties from the "Divide California into ... " navigation on the left, and then click to download data at the top right. One of the choices is actually a shapefile including both geography and data. Download that, unzip it, and import it with

```
ca_income <- read_shape(
  "data/path2file/acs2015_5yr_B19013_05000US06063.shp",
  as.sf = TRUE)
```

(substitute the path to your unzipped shapefile and the name of your shapefile).

If you are interested in seeing the shapefile's projection and datum, import it with the sf package's st_read() function instead of read_shape(). For example:

```
ca_income <- sf::st_read("data/path2file/acs2015_5yr_B19013_05000US06063.shp").
```

```
## Reading layer `acs2015_5yr_B19013_05000US06063' from data source `D:\Sharon\My
## Documents Data Drive\BookMarkdown\data\acs2015_5yr_B19013_05000US06063\acs2015_5yr_
## B19013_05000US06063\acs2015_5yr_B19013_05000US06063.shp' using driver `ESRI Shapefile'
## Simple feature collection with 59 features and 4 fields
## geometry type:  MULTIPOLYGON
## dimension:      XY
```

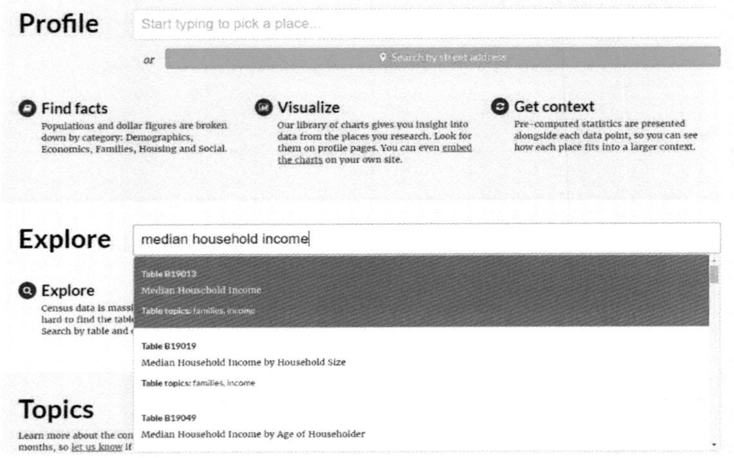

Figure 11.3: CensusReporter.org interface

```
## bbox:           xmin: -124.482 ymin: 32.52883 xmax: -114.1312 ymax: 42.00952
## epsg (SRID):    4326
## proj4string:    +proj=longlat +datum=WGS84 +no_defs
```

Run `str(ca_income)`, and you'll see that the county column is called "name," the median income column is called "B19013001," the margin of error is "B19013001e", and spatial information is in a special list column called "geometry". You don't want to touch the geoid or geometry column names, but you can change the names of columns 2 through 4 with

```
names(ca_income)[2:4] <- c("County", "Median.Income", "MoE")
```

or explicitly rename them with dplyr's rename():

```
ca_income <- rename(ca_income, County = name, Median.Income = B19013001, MoE = B19013001e)
```

Before mapping, take a look at the income distributions with a histogram in base R to get a feel for the data (see Figure 11.4). `options(scipen = 999)` will get rid of scientific notation along the x axis.

```
options(scipen = 999)
hist(ca_income$Median.Income)
```

It's easy to see a basic map of the data with tmap's qtm() function. Tell qtm() which data column to use to color by, with the syntax `qtm(mygeodataframe, fill = "colname")`, such as:

```
qtm(ca_income, fill = "Median.Income")
```

Just like that, you've got a quick data visualization of median income by county on a map (Figure 11.5). (Note: It's possible to add labels to the polygons with a `text = "colname"` argument, but the map would need to be larger than available space in this book's print version.)

11.5 An even easier way to pull U.S. Census data

Kyle Walker's newer tidycensus package feels almost like cheating, it's so easy: You can import Census data already combined with geometry with just a couple of commands, all in R-ready format.

You can store your Census API key for tidycensus to use by running `census_api_key("YourCensusKey")`. Then, run a line of code such as `ca_income <- get_acs(state = "CA", geography = "county",`

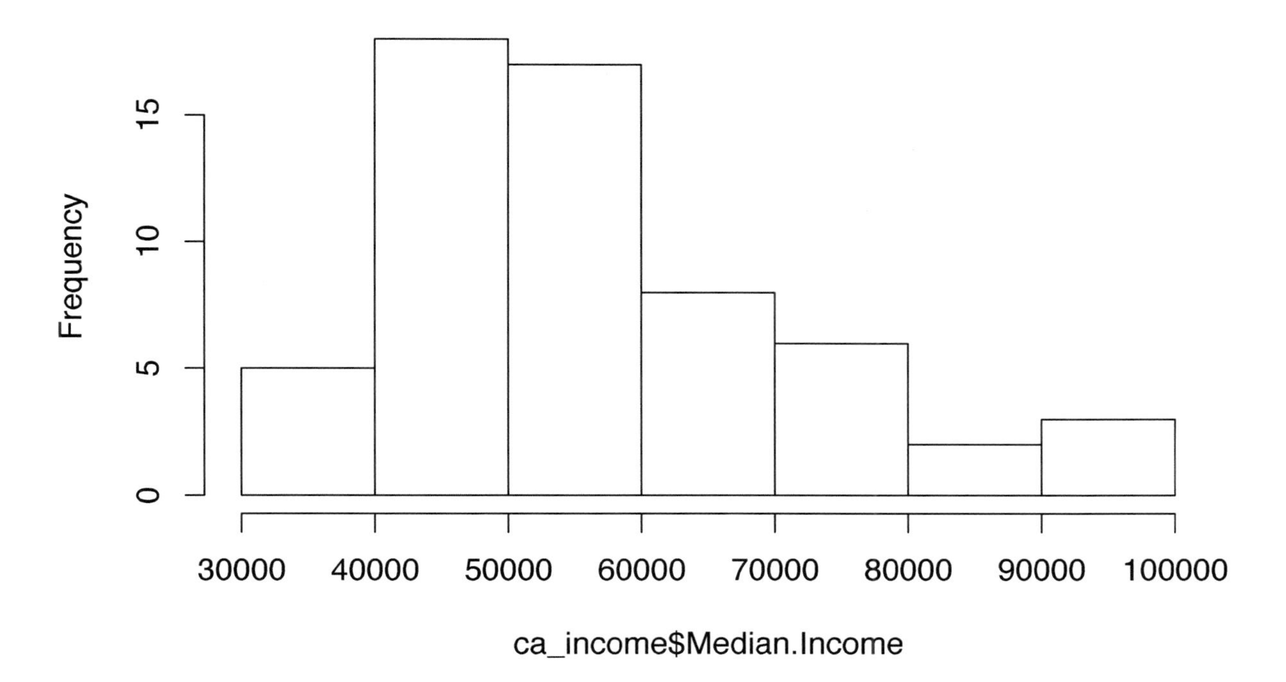

Figure 11.4: Histogram of California median income by county.

`variables = "B19013_001", geometry = TRUE)`. You'll get a simple-features object in return, which can be easily mapped using tmap, ggplot2, or other data visualization packages that can map sf objects.

As of this writing, tidycensus supports the recent 10-year census and five-year American Community Survey. See https://walkerke.github.io/tidycensus/ for more information about the package. Walker also has some helpful blog posts such as Compare US metropolitan area characteristics in R with tidycensus and tigris, at https://walkerke.github.io/2017/06/comparing-metros/.

11.6 Interactive maps with tmap

It would be nice to be able to click or hover over a county and get all the underlying data while exploring the map. That's possible with tmap.

tmap has two display modes: the default, "plot", which creates a static map; and, more recently added, an interactive mode called "view." To change modes, use the tmap_mode() function: `tmap_mode("view")`.

If you run `qtm(ca_income, fill = "Median.Income")` again and are patient – it may take a little while to finish – you'll see you can click on a county and get the specific *median income* number. Unfortunately, you won't see the *name of the county.* Unfortunately, as far as I know, adding columns to the pop-up window isn't supported in qtm(), but you can do this with tmap's full-featured mapping function. qtm() would let you label the counties on the map itself with text = "County", but there's not enough room on the map to fit all the county names.

There's a useful shortcut for the previous two steps of switching tmap mode and re-drawing the most recent map. tmap's ttm() function toggles between the two views, while tmap_last() redraws the prior map. So,

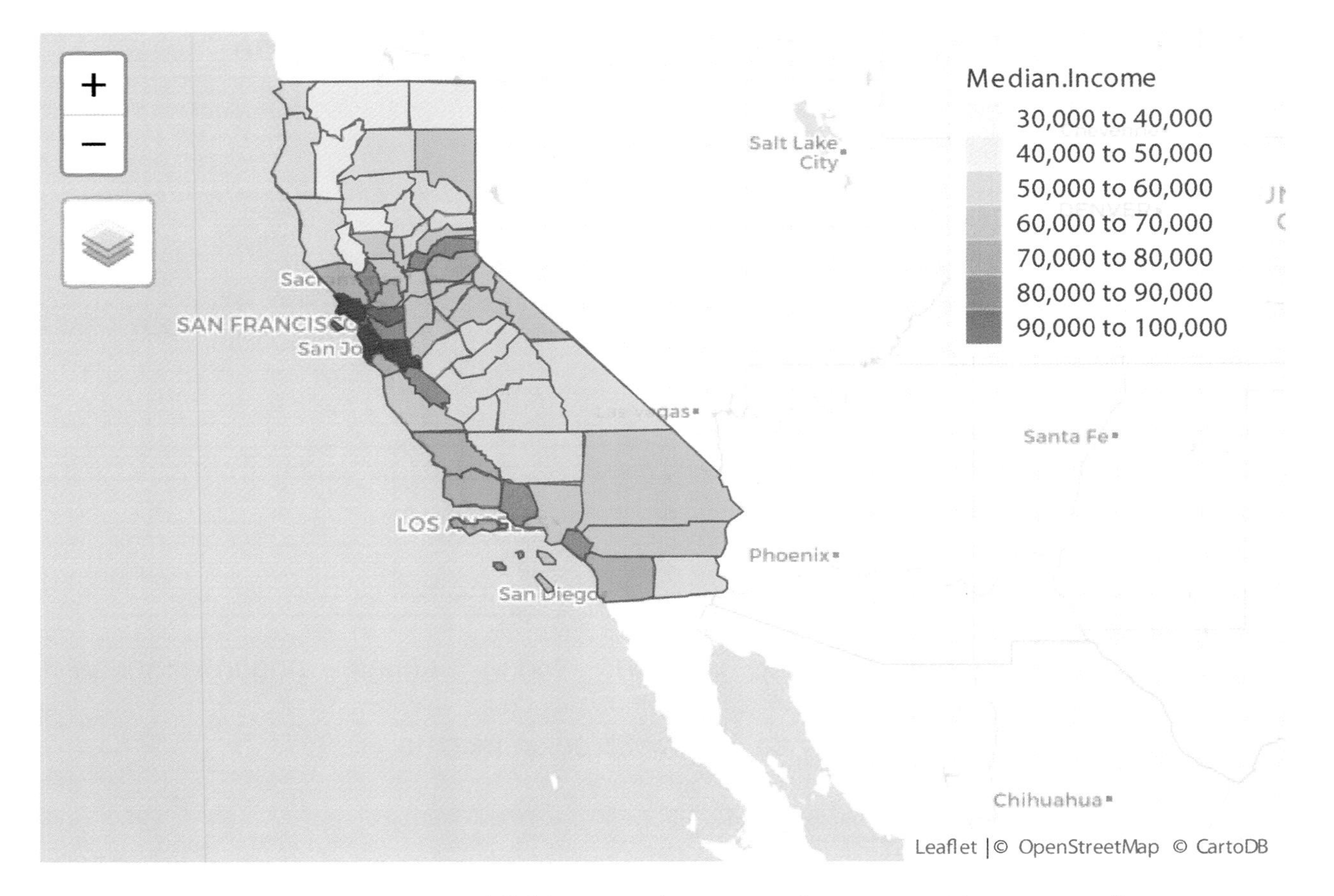

Figure 11.5: Easy map of California median income by county with tmap::qtm().

```
ttm()
tmap_last()
```

would have switched you from plot to view mode and redrawn the map in interactive mode.

tm_shape() and tm_polygons() are tmap's more robust mapping functions for choropleth maps. This is how to create the county map with county labels in the pop-up window:

```
tm_shape(ca_income) +
  tm_polygons(col = "Median.Income", id = "County")
```

Click on a county now and you'll get the county name as well as median income.

Like ggplot2, this map is created with layers (although the layering format is somewhat different). Let's take a look at how this works.

tm_shape() in the first line of code defines what object has the data. tm_polygons() adds a polygon layer and controls how the polygons should display. The col argument says which column should be used for the polygons' colors; id sets the title of the pop-up window. (The popup.vars argument to tm_polygons would let you add additional variables to the pop-up window.)

You can look at this map in RStudio's Viewer pane in the bottom right, pop open a larger window with that pane's Zoom icon, or open it in a browser by clicking on the "Show in new window" icon (the icon to the right of the broom). In addition, if you store your map in an object while in view mode, such as

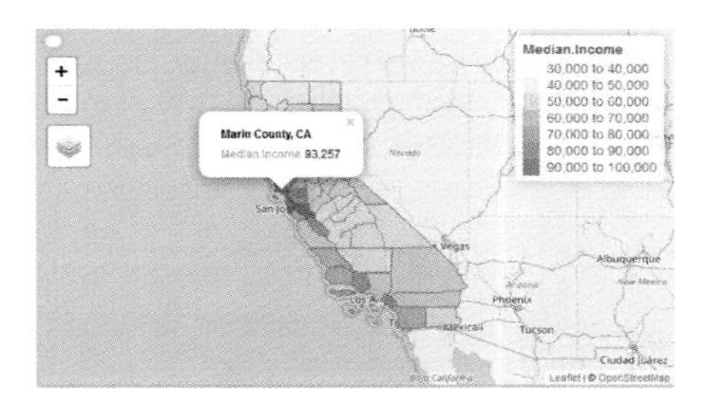

Figure 11.6: Interactive maps with tmap can include rollovers and popups.

```
ca_income_map <- tm_shape(ca_income) +
  tm_polygons(col = "Median.Income", id = "County")
```

you can use the tmap_save() function to save the interactive map to a self-contained html page:

```
tmap_save(ca_income_map, "CA_Counties_Map.html")
```

It's then possible to post that page online, either stand-alone or in an iframe on another page.

There's a lot more tmap can do. You can add additional polygon or point layers one by one, similar to building a visualization with ggplot2. tmap also supports customizing colors with the palette argument. For example, in a tm_polygons() layer, you would use a syntax such as: `tm_polygons(col = "Median.Income", id = "County", palette = "Greens")`.

What color scheme to use? The tmaptools package has a great way to explore available Color Brewer palettes, allowing you to adjust for several different factors such as number of colors (for categorical variables) and color blindness. Try it by running `palette_explorer()` and you should see a window open that looks something like Figure 11.7.

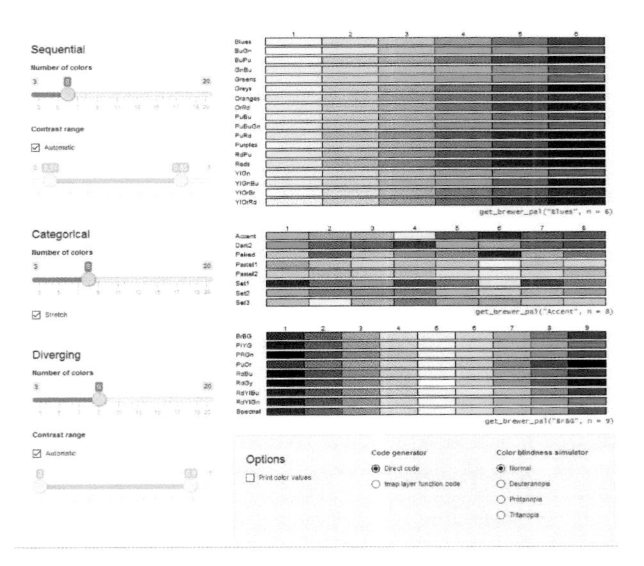

Figure 11.7: tmaptools::palette_explorer() can help you choose a color scheme for your maps.

If this doesn't work on your system, you may first need to manually install the shiny and shiny.js packages.

Sequential palettes are for cases like median income, where you've got numerical variables going from low to high. *Categorical* palettes are for non-numerical categories such as gender or political party. *Diverging* is when you've got data with a center and values going out in two directions, such as looking at whether housing prices have increased or decreased.

Once you find a palette you like, you can use it by referring to its name in the palette argument, such as palette = "Greens" above. In addition to choosing a Color Brewer palette, you can also decide how many different colors you want the palette to have, with the n argument, such as `palette = "Greys", n = 8`. Complete code:

```
ca_income_map <- tm_shape(ca_income) +
  tm_polygons(col = "Median.Income", id = "County", palette = "Greys", n=8)
```

Take a look at the "tmap in a nutshell"" vignette for more information (if it's not available on your system, look for it on the package's GitHub repository.)

11.7 Importing and joining data

Data you'd like to map and analyze isn't always conveniently available already packaged in a single shapefile, of course. Often, geography will be in one file, numerical data will be in another file, and it will be your job to merge the two.

That's the case if we want to map California unemployment data by county. The shapefile comes from the Census Bureau, while unemployment rates are from the U.S. Bureau of Labor Statistics.

Here's how to make a map from both sets of data.

1. *Import the shapefile* of California county geography as we did before. Since the shapefile is available from the U.S. Census bureau via the tigris package, `ca_counties <- counties("CA", class = "sf")` will work. If you were using a shapefile from another source, you could import it with `tmaptools:read_shape("shapefile.shp", as.sf = TRUE)` or `read_sf("myshapefile.shp")`.
2. *Import the data* you want to map. You can download data from the BLS or use my sample data from the book GitHub repo at *data/CA_unemployment.xlsx.*

The download.file() function will need the `mode = "wb"` argument so that the file is downloaded *as a binary file.* That's needed for this Excel file. It wouldn't be needed for a plain-text file, like a CSV. Basically, any file that you can't open and read with a text editor like Notepad on Windows or TextWrangler on Mac should be downloaded as a binary file.

rio::import() will needs the skip = 1 argument, because details about the data are in the first row of the spreadsheet; actual data with column headers starts on row 2. quiet = TRUE is optional; it just stops download-progress messaging.

Here's the code to download the data from BLS that I posted on my personal site:

```
download.file("http://www.machlis.com/SampleData/CA_Unemployment_2016-12.xlsx",
    destfile = "CA_unemployment.xlsx", mode = "wb", quiet = TRUE)
ca_unemployment <- rio::import("CA_Unemployment.xlsx", skip = 1)
```

Or, load it from the files you downloaded from this book's GitHub repository with `ca_unemployment <- rio::import("CA_Unemployment.xlsx", skip = 1)`.

3. *Join the data by common column.* There are two things you need to check before doing a join: which column in each data frame holds the common key, and whether each county's name is *exactly the same* in both of those columns.

This is important. "Sierra" is not the same as "SIERRA" or "Sierra County". And even if most listings look the same, sometimes you can get tripped up with something like "North Adams" in one data set and "N. Adams" in another.

The ca_unemployment data frame lists county names like Alameda County and not simply Alameda. So, I looked for a similar structure in the ca_counties data and found the NAMELSAD column, which also has "County" after the county name.

After finding the common columns, I calculated **intersection** – items that appear in both columns – with R's union() function. There should be as many *common* counties in both data sets as there are *total* counties. You can eyeball this to check number of items in the intersection and number of rows in the county data frame. Or, you can write a little code to test whether the length of the intersection (common_counties) is the same as the length of all the counties in the shapefile (`ca_counties$NAMELSAD`). The == function tests for equality:

```
common_counties <- union(ca_counties$NAMELSAD, ca_unemployment$County)
length(common_counties) == length(ca_counties$NAMELSAD)
```

```
## [1] TRUE
```

I don't really like NAMELSAD as a column name, so for this map, I'll change it to County in ca_counties, using dplyr's rename() function. That was covered in Chapter 6, but a reminder that the syntax is `rename(newColumnName = oldColumnName)`:

```
ca_counties <- dplyr::rename(ca_counties, County = NAMELSAD)
```

dplyr functions like rename() work with simple features objects but not SpatialPolygonsDataFrames – another reason I prefer working with simple features. Now it's time to combine the two data sets. Adding data to a shapefile is slightly different from merging two plain data frames. Instead of using merge() or left_join(), I like tmaptools' `append_data()` for combining numerical data with geospatial data. The format is `append_data(shape_object, data_object, key.shp = shapeColumnKeyName, key.data = dataColumnKeyName)`:

```
ca_joined <- tmaptools::append_data(ca_counties, ca_unemployment,
                key.shp = "County", key.data = "County")
```

```
## Keys match perfectly.
```

If you didn't change the ca_counties$NAMELSAD column name, key.shp would equal "NAMELSAD" instead of "County".

4. *Map as usual (see Figure 11.8).*

```
tm_shape(ca_joined) +
  tm_polygons(col = "Rate", id = "County", palette = "Oranges")
```

11.8 Leaflet and points on a map

11.8.1 Project: Mapping bank locations

Why might you want to map bank locations? Maybe activists have argued that financial institutions are avoiding areas with large minority populations. Or, perhaps citizens in a historic neighborhood are unhappy that a cherished local store is being taken over by a new branch of a national bank, arguing that they're already overrun with banks compared to nearby areas.

In general, learning how to place locations on an interactive map can come in handy for a wide variety of stories, from showing local polling places to mapping marijuana dispensaries.

The leaflet package, created by RStudio, is like ggplot2 and tmap in that it also produces visualizations by layers. You start off defining a map object; add map background tiles; set the latitude, longitude and zoom level for how the background map should display; and then layer on additional features such as points and polygons, legends, and labels. We'll go through all these steps to create a full-featured map.

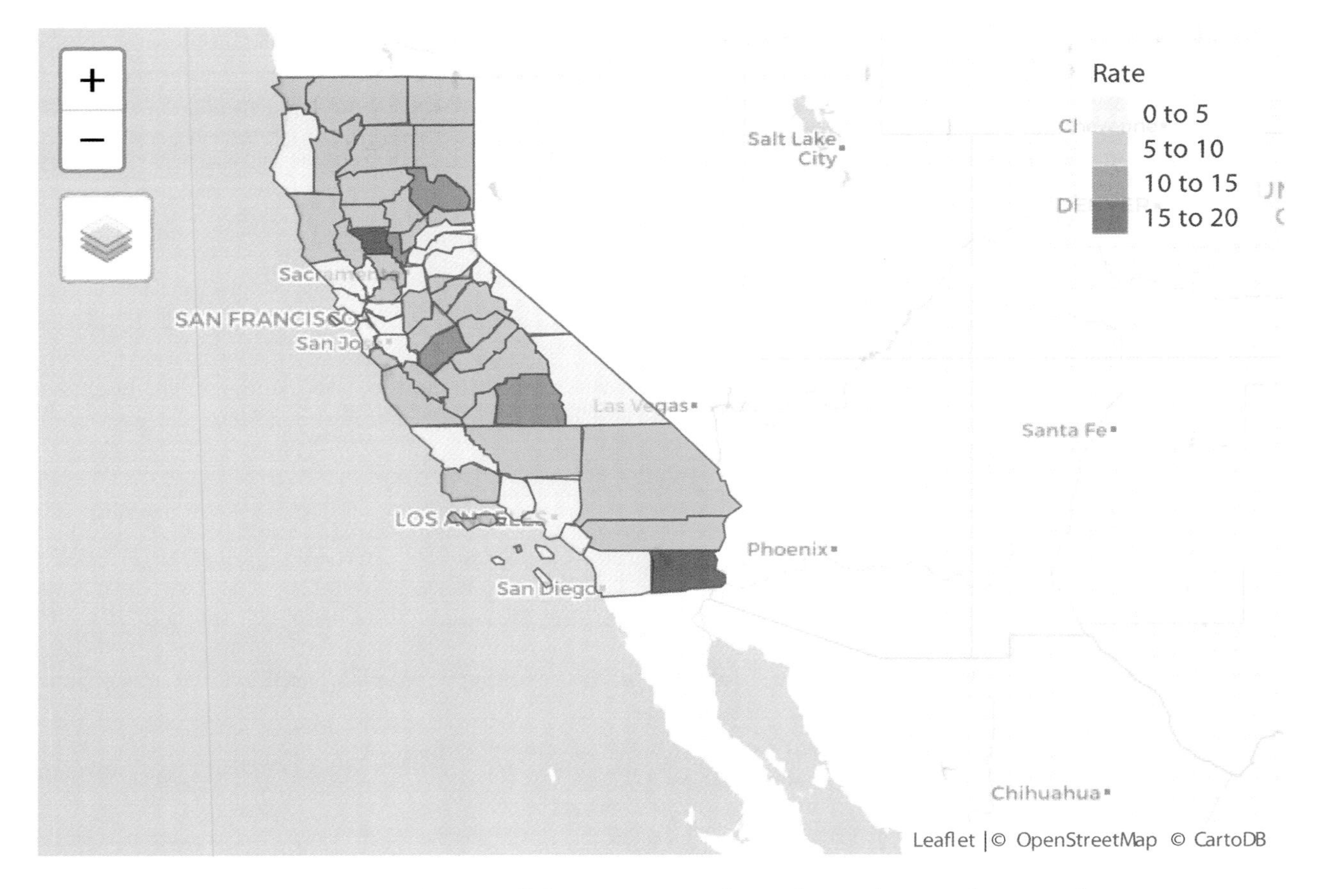

Figure 11.8: Map of joined California geography and joined unemployment data.

11.8.2 Geocoding with ggmap

Leaflet needs latitudes and longitudes in order to place points on a map, not just addresses. Chances are, your project will have a list of addresses or intersections and not latitude and longitude. There are a number of geocoding services that will generate lat/lon data for you. One of the most convenient is to geocode right within R using the ggmap package.

If you had loaded ggmap the conventional base R way with `library(ggmap)` instead of pacman::p_load(), you'd see a link to Google Maps API Terms of Service and a request to cite the package if you use it. Keep those in mind for any projects you plan to publish. p_load() loads packages quietly, supressing start-up messages.

Now, look at the help file for ggmap's geocode() function using `?geocode`. You'll see that the only input the function requires is a character vector of addresses. Note that they need to be *characters.* If you have a data frame with an address column that came in as *factors* instead of *characters*, you'll first need to convert that column to characters with the as.character() function.

The project data will be a CSV file of banks in Boston, created from the U.S. Federal Deposit Insurance Corp.'s list of federally insured bank locations combined with information about state-chartered and -insured institutions from the Massachusetts Division of Banks.

That merged federal and state file is in the book's GitHub repo you downloaded, at *data/BostonBanks.csv.*

11.9 geocoding and R's paste() function

Import the file and take a glimpse at the structure.

```
bosbanks <- rio::import("data/BostonBanks.csv")
str(bosbanks)
```

```
## 'data.frame':    289 obs. of  6 variables:
##  $ Address: chr  "1 Lincoln St." "280 Congress Street" "50 Rowes Wharf" "1
## Boston Pl" ...
##  $ County : chr  "Suffolk" "Suffolk" "Suffolk" "Suffolk" ...
##  $ City   : chr  "Boston" "Boston" "Boston" "Boston" ...
##  $ State  : chr  "MA" "MA" "MA" "MA" ...
##  $ Zip    : int  2111 2210 2110 2108 2110 2134 2110 2109 2122 2114 ...
##  $ Bank   : chr  "State Street Bank And Trust Company" "Manufacturers And Traders
##  Trust Company" "Jpmorgan Chase Bank, National Association" "The Bank Of New York
## Mellon" ...
```

Right now each address is *scattered across several columns:* Address (which is just the street address), City, and State. According to ggmap's geocode() help file, the address to be geocded should be a single character string. So, we need to create a new address column from several existing columns, with a format of `Address, City, State`.

R's paste() function combines two or more character strings into one, with a default of a space separating each one. That seems promising, except we need a comma and space separating each item, not just a space. Fortunately, you can change what separates the character strings you're joining.

`paste("Boston", "MA")` returns `Boston MA`. You can create one `Boston, MA` string from "Boston" and "MA" with paste by changing the separation string from a space to a comma and a space: `paste("Boston", "MA", sep = ", ")`.

There's also a variation of paste(): paste0(), which is paste() without any separating character. The basic syntax is `paste0(string1, string2, string3)` and so on. `paste0("Boston", ", ", "MA")` concatenates Boston, a comma and space, and MA to produce `Boston, MA`. Sometimes paste0() is easier and more intuitive than paste(), but this doesn't seem to be one of those times. So I'll stick with paste().

Can you create the string `"271 Great Road, Acton MA"` from `"271 Great Road"` `"Acton"` and `"MA"`? That's a good hint about how to get the address column we need.

Here's one way to create a new address column, CompleteAddress, from the Address, City, and State columns:

```
bosbanks$CompleteAddress <- paste(bosbanks$Address, bosbanks$City, bosbanks$State, sep = ", ")
```

Another way: Use dplyr's mutate function to create a new column in the data frame (discussed in Chapter 8). This is a slightly more elegant and less repetitive version:

```
bosbanks <- bosbanks %>%
  mutate(
    CompleteAddress = paste(Address, City, State, sep = ", ")
  )
```

A paste() alternative: The glue package. Pasting more than a few strings and variables together can start getting cumbersome, though. The glue package's glue() function will return the *value of variables enclosed in {} braces* within a quoted string. In other words, anything inside brackets is evaluated; everything else is returned as is.

If the value of x was 3, `glue("My variable called x is equal to {x}.")` would return the sentence `My variable called x is equal to 3."` So, `glue("{Address}, {City}, {State}")` is similar to `paste(Address, City, State, sep = ", ")` or `paste0(Address, ", ", City, ", ", State)`. Which you use depends entirely on the syntax you like most.

11.10 Time to geocode with R (or maybe without)

ggmap has its own geocoding function that mutates, mutate_geocode(). It geocodes an address column and adds appropriate latitude and longitude functions to a data frame all in one step. Very convenient! It's as easy as `mydata <- mutate_geocode(mydata, AddressColumnName)`.

In theory.

```
bosbanks_geocoded <- mutate_geocode(bosbanks, CompleteAddress)
```

Note: With close to 300 addresses, this will take a little while to run. You may want to try this with a subset of bank data, such as

```
sampledata <- bosbanks[1:10,]
sampledata <- mutate_geocode(sampledata, CompleteAddress)
```

and then import my geocoded file with `bosbanks <- rio::import("")`. If you do geocode the file yourself, make sure to save it to a CSV file so you won't have to do that again (reminder: `rio::export(bosbanks, "BostonBanks.csv")` will do that)

The Google geocoding API has a limit of 2,500 calls per day. If you're not in a rush and your file isn't that much over 2,500, you can break it up into pieces to run over several days. Unfortunately, because the package doesn't ask for an API key, I've often run into limits even when geocoding far fewer than 2,500.

Another option for geocoding is to use the open-source Data Science Toolkit instead of the Google Maps API for geocoding. You can specify that ggmap use the Data Science Toolkit with `sampledata <- mutate_geocode(sampledata, CompleteAddress, source = "dsk")` . More info about the Toolkit itself at http://www.datasciencetoolkit.org/.

There are other choices besides ggmap. One uses the Geocodio service, which allows up to 2,500 free lookups per day (50 cents per 1,000 after that). Bob Rudis created an R package for Geocodio at https://github.com/hrbrmstr/rgeocodio. Apply for a free Geocodio key at geocodio.

After installing and loading rgeocodio from GitHub with `pacman::p_load_gh("hrbrmstr/rgeocodio")`, you can batch geocode the complete address column with

```
bosbanks_geocodio <- gio_batch_geocode(bosbanks$CompleteAddress)
```

The problem for beginners is that gio_batch_geocode() returns latitude-longitude results as a *complex list column within a data frame*, not as simple, separate columns.

It's a nice package and service, though. If you'd like to try it, you should be able to extract lat and lon data from the response_results list column using purrr's map_dbl() function (to create a vector of numbers with decimal points)

```
bosbanks <- bosbanks_geocodio %>%
  mutate(
    lat =  purrr::map_dbl(bosbanks_geocodio$response_results, ~.$location.lat[1]),
    lon = purrr::map_dbl(bosbanks_geocodio$response_results, ~.$location.lng[1])
  )
```

Honestly, though, although this is an R book, if you're only dealing with data like this occasionally and don't *really* need the geocoding as part of a reproducible workflow, you can just head over to https://geocod.io and upload a spreadsheet or CSV of less than 2,500 addresses manually. You won't even need to combine street address, city, and state into a single column, since the Web service understands properly named columns.

One final data-cleaning note: Before I cleaned the files, some addresses included text such as "23rd floor" within the street address column. rgeocodio was smart enough to understand that syntax, but ggmap wasn't and some geocoding failed. I removed any floor information when cleaning and merging the data.

You can read in an already-geocoded version of the data file using the files you downloaded from this book's GitHub repo with:

```
bosbanks <- rio::import("data/BostonBanks_geocoded.csv")
```

11.11 Mapping points with leaflet

We're now finally ready to map Boston bank branch locations as points with the leaflet pakage. This will be slightly different than mapping counties as polygons. We'll need to do the following:

Step 1: Create a *map object* with the `leaflet()` function and save it to a variable (here I called it bosbankmap). Don't forget to load the leaflet package with `pacman::p_load(leaflet)` or `library(leaflet)` (first running `install.packages("leaflet")` if you didn't run the script at the beginning of the chapter for packages needed.)

```
bosbankmap <- leaflet(data = bosbanks)
```

Nothing will happen yet beyond getting a blank canvas (Figure 11.9), which you can see by typing `bosbankmap` in your console:

Figure 11.9: A plain leaflet() object shows a blank canvas.

Step 2: Add background tiles. The default addTiles() will put Open Street Map tiles on your map.

```
bosbankmap <- bosbankmap %>%
  addTiles()
```

If you're not familiar with map tiles, interactive Web maps like Google Maps use them. That's what lets you see a small portion of the world and then other nearby areas as you click and drag or zoom in and out. This will probably make more sense if you run some code to see it for yourself.

A credit to OpenStreetMap, license info, and link back to openstreetmap.org are required for Web usage. Fortunately, the leaflet package takes care of that for you, as in Figure 11.10.

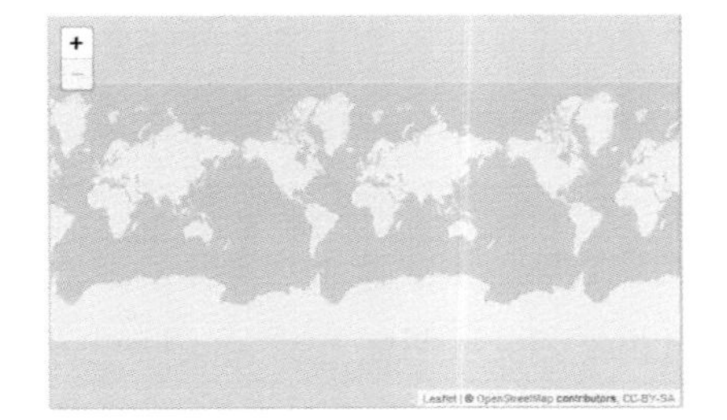

Figure 11.10: Map tiles added to a leaflet() object.

There are other map tile options available in leaflet. Dozens are built in; see the options with `names(providers)`. Some, such as Esri and Mapbox, require you to register for an account. Also make sure to check out terms of service, since some aren't open source.

To use another provider, the syntax is `addProviderTiles(providers$FreeMapSK)`.

Step 3: Set where you want the map to be centered as well as its zoom level with leaflet's setView(), using the syntax `setView(mapObject, longitude, latitude, zoomLevel)`. Since this is a map of Boston, I want the map centered there. ggmap can give me general longitude and latitude information:

```
ggmap::geocode("Boston, MA")

## Information from URL : http://maps.googleapis.com/maps/api/geocode/json?address=Boston,
## %20MA&sensor=false

##          lon      lat
## 1 -71.05888 42.36008
```

I can't give you much advice on how to set the zoom level, except that it's pretty much a try-and-tweak operation. I'll start with 10 (Figure 11.11) and see what that looks like.

```
bosbankmap %>% setView(-71.05888, 42.36008, zoom = 10)
```

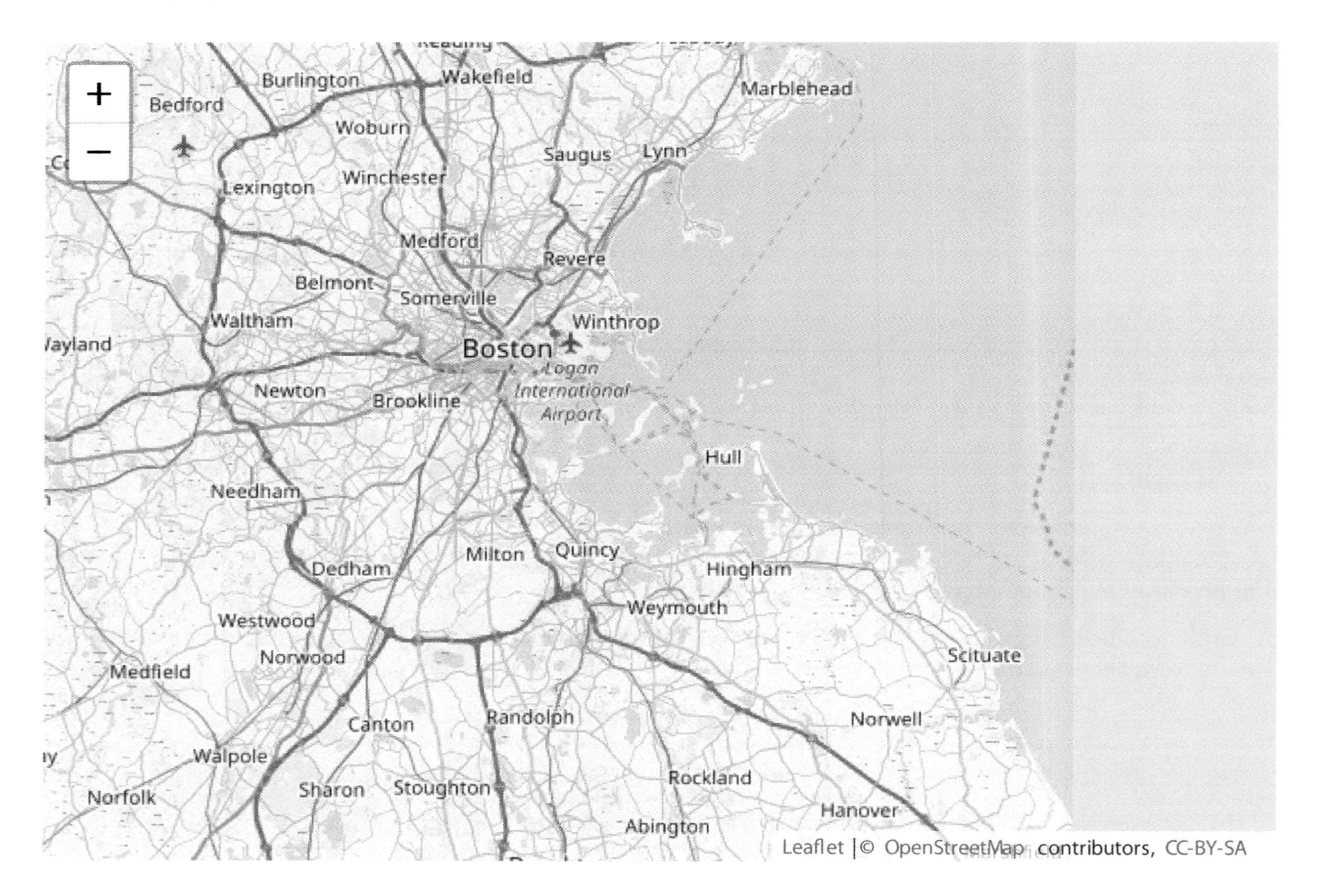

Figure 11.11: Map at zoom level 10.

Not zoomed in enough, so I'll raise the zoom a few (Figure 11.12):

```
bosbankmap %>% setView(-71.05888, 42.36008, zoom = 14)
```

That seems OK to start, so I'll add that to the bosbankmap object:

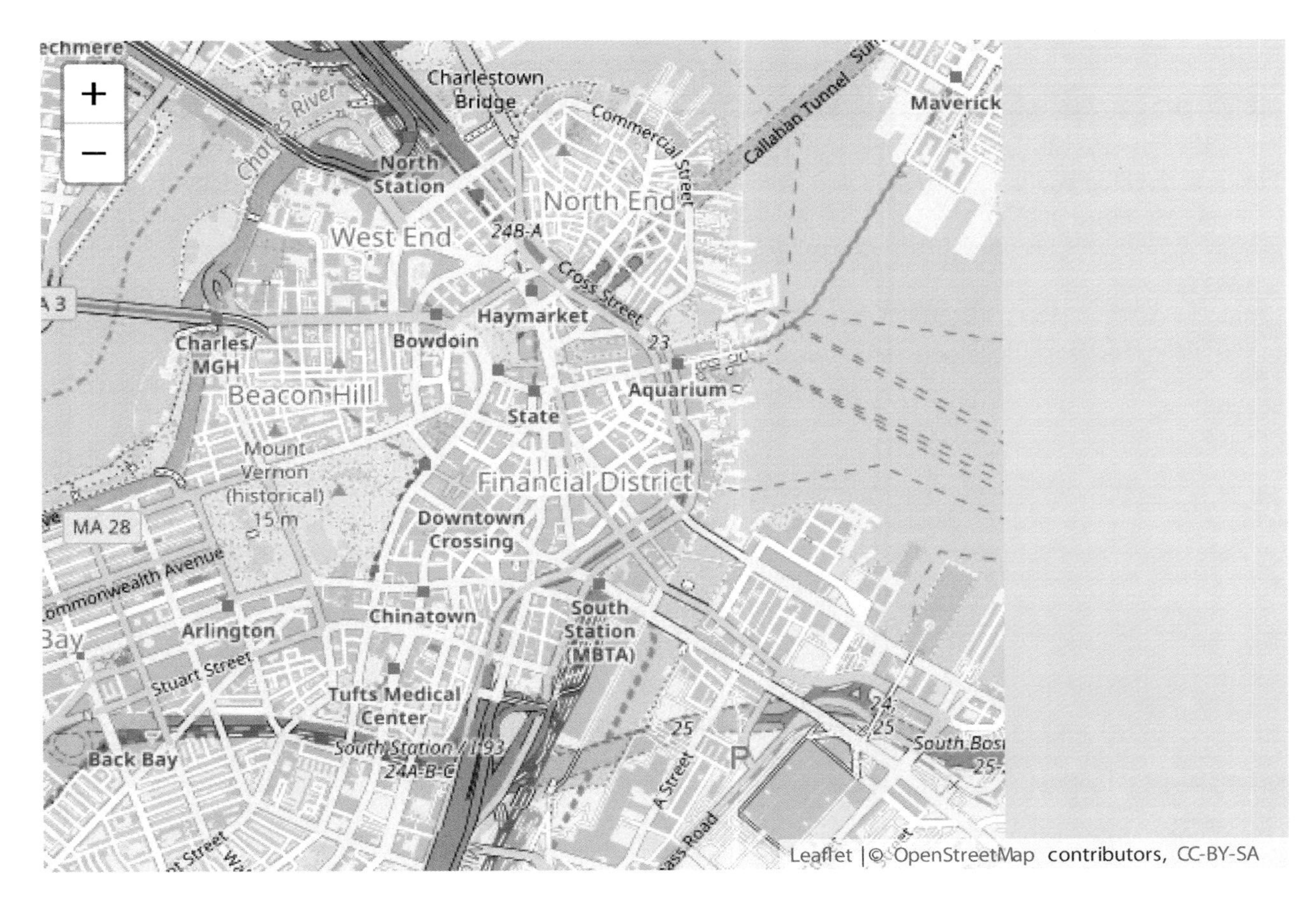

Figure 11.12: Map at zoom level 14.

```
bosbankmap <- bosbankmap %>%
  setView(-71.05888, 42.36008, zoom = 14)
```

Step 4: Add points to the map with the addMarkers() function, using the following arguments: `addMarkers(map = mymapobject, data = mydataframe, lng = ~LongitudeColumnName, lat = ~LatitudeColumnName, popup = ~PopupColumnName)`.

Hopefully, the data, latitude, and longitude arguments are pretty self-explanatory. We can explicitly state ~lon and ~lat for the longitude and latitude column names, but we don't have to, because leaflet's addMarkers() function will assume a column named lon (or lng, long, or longitude) is for longitude, and lat or latitude is the latitude. We also don't need to use `data = bosbanks` with addMarkers(), since we already set the data source in line 1 (`leaflet(data = bosbanks)`).

```
bosbankmap <- bosbankmap %>%
  addMarkers(data = bosbanks, popup = ~Bank)
```

```
## Assuming "Longitude" and "Latitude" are longitude and latitude, respectively
```

Type the name of the map, bosbankmap, into your console, wait for it to render, and click on a marker. You should see the bank name.

It would be more interesting to design our own pop-up text, though. Leaflet pop-ups can display HTML. And that means we can create a new column including HTML tags such as for bold and line breaks. For example,

```
bosbanks$popuptext <- paste0("<b>", bosbanks$Bank, "</b><br />",
                    bosbanks$Address, "<br />", bosbanks$City)
```

will generate a text column with a format like "<b>State Street Bank And Trust Company</b>
1 Lincoln St. Fl 1
Boston".

This is something you'll want to do before your leaflet code reads in the data frame holding data. If you want to design your own pop-up text, do so; re-create the leaflet map; and add addMarkers(popup = ~popuptext):

```
bosbankmap <- leaflet(data = bosbanks) %>%
  addTiles() %>%
  setView(-71.05888, 42.36008, zoom = 14) %>%
  addMarkers(popup = ~popuptext)

bosbankmap
```

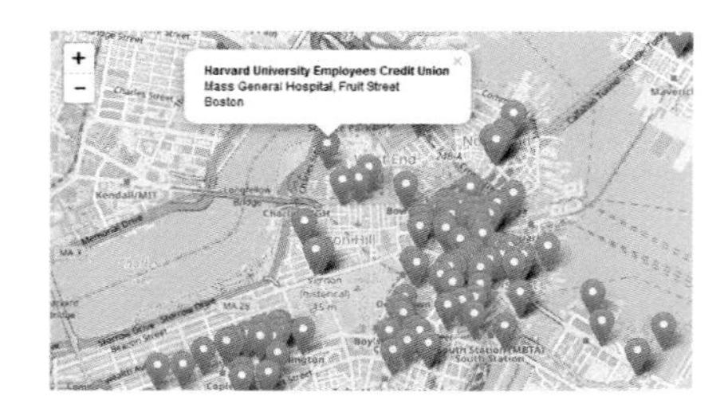

Figure 11.13: leaflet map with pop-ups markers.

Now let's look at the map (Figure 11.13)!

One immediately visible gap is in Boston's Beacon Hill area around the State House. This is an extremely expensive neighborhood, where avoiding minority/low-income areas is definitely *not* a factor. The gap here is likely zoning – Beacon Hill is a historic area with very stringent rules about what can be built or changed. That brings up another point before jumping to conclusions about where banks do and don't *choose* to locate: It would be important to look at a zoning map to see where banks aren't permitted.

Zoom out, and you see another gap – and this one is in a minority neighborhood. But it's also around 557-acre Franklin Park. Is that the reason, or is something else going on? More research and reporting would be needed before jumping to the conclusion that banks are actively avoiding a neighborhood if you spot a gap. What are the zoning rules? Maybe you could get a zoning map and overlay it on this one.

What do community leaders say about the state of neighborhood banking? Have they been trying to attract banks to serve the neighborhood, or is the presence of ATMs (not mapped) more than adequate? Has the economic development focus been on, say, trying to get a grocery store instead – in which case, it might be a better project to get a list of all groceries in Boston and map those. If you decided to switch this project to grocery stores, the hardest part will likely be getting that list of grocery stores and addresses. Now that we have the map code, it would be fairly trivial to swap in another data set and re-produce the map with new data points.

One last point about the points: I didn't check for multiple points at a single location. For this example, I was looking for gaps, not concentrations, so it wasn't that important. If you wanted to look for clusters, though, the markercluster plugin might be useful. It can show how many numbers are being represented by a single marker. There's more about that on the RStudio leaflet package documentation at https://rstudio.github.io/leaflet/markers.html.

11.12 Points and polygons on a single map

It can be interesting to combine the two types of maps we've just done, and see *points mapped within different-colored polygons* such as neighborhoods, precincts, or U.S. Census tracts. This kind of map could help illustrate a lot of different stories or policy reports. Where are polling locations? Fire stations? Marijuana dispensaries? Schools? Incinerators?

These kinds of maps can be designed simply to give information about locations – here's the polling place in each precinct. Or they can help with analysis, such as whether there might be a pattern between, say, socioeconomic factors and placement of things like incinerators in a state.

For this section I'll stay with banks in Boston neighborhoods, since we've already prepped the data. The City of Boston has a neighborhood shapefile available on its data portal at data.boston.gov that I included in this book's GitHub repo files. You can unzip it with R and then import it by changing to the data directory, unzipping with unzip(), and then importing the shapefile with read_shape():

```
setwd("data")
unzip("Boston_Neighborhoods.zip")
bos_geo <- read_shape("Boston_Neighborhoods.shp", as.sf = TRUE, stringsAsFactors = FALSE)
setwd("..")
```

As I've mentioned before, I find the tm package easiest for interactive choropleth maps and leaflet best for interactive points on a map. Fortunately, it's easy to turn tm-created maps into leaflet objects, so I can use tm for polygons, leaflet for points, and combine the two.

Here's how:

```
# Make sure tm is in interactive mode
tmap_mode("view")

## tmap mode set to interactive viewing
# Map the neighborhood polygons in tmap and then turn the result into a leaflet object:
mypolygons <- tm_shape(bos_geo) +
  tm_polygons(col ="Name", id = "Name", alpha = 0.7)
mypolymap <- tmap_leaflet(mypolygons)

# Use the leaflet package's addMarkers() function to add points to the choropleth map:
mypolymap %>%
  addMarkers(bosbanks$Longitude, bosbanks$Latitude, popup = bosbanks$popuptext)
```

If you're not going to remember how to do this but expect to create maps like this, you can always *make a code snippet from this code!*

Finally, if you're looking to see whether certain points are clustered or barren in minority neighborhoods, you're probably better off overlaying racial demographic data than neighborhood names, since it's possible socioeconomic status varies within neighborhoods (especially in areas that are rapidly changing). In the U.S., you can find such data from the Census Bureau.

I downloaded a shapefile of Boston racial data from the Census Reporter site by searching for place Boston and table B02001 and then downloading the resulting shapefile to the data subdirectory.

The code below imports it into R. I created a new PctWhite column by dividing the white population in column B02001002 / the total population in column B02001001. Finally, I deleted the last row, because it is a summary total row, using dplyr's slice() function. In addition to using slice to define what rows you want to *keep*, you can also use it to select what rows to delete with a minus sign. `slice(mydf, -n())` removes the last row of mydf (n() is the total number of rows in an object within a dplyr pipe analysis).

```
bosrace_geo <- read_shape("data/BostonRacial/acs2015_5yr_B02001_14000US25025090600.shp",
      as.sf = TRUE, stringsAsFactors = FALSE) %>%
  mutate(
    PctWhite = round((B02001002 / B02001001) * 100, 1)
  ) %>%
  slice(-n())
```

The next section of code is similar to the previous points-and-polygons map, but with a couple of customizations in the tm_polygons() function. `col="PctWhite"` we've already seen (or something very similar). That says we want the PctWhite column values to be the ones that control the map's color scale.

alpha says how transparent or opaque the coloring should be, using a scale from 0 to 1: 0 is completely transparent while 1 is opaque.

n says how many different colors I want on the map. The default is 5, which would give 20 percentage points in each category from 0% to 100%; but I decided to manually override that to create 10 categories of 10 percentage points each.

palette sets the particular color scheme we want to use, with a couple of extra customizations. get_brewer_pal() generates a specific Color Brewer palette – "YlOrBr" says I want the yellow-to-orange-to brown palette, and 10 says I want ten colors in my palette. The rev() function around get_brewer_pal() means I want the palette to be *in reverse of its usual order*, so the deeper colors are for lower numbers and lighter colors are for higher numbers.

Here's the full code:

```
mypolymap <- tm_shape(bosrace_geo) +
  tm_polygons(col ="PctWhite", alpha = 0.7, n = 10,
              palette = rev(get_brewer_pal("YlOrBr", n=10)))
mypolymap <- tmap_leaflet(mypolymap)

 mypolymap %>%
  addMarkers(bosbanks$Longitude, bosbanks$Latitude, popup = bosbanks$popuptext)
```

Figure 11.14: Mapping points and polygons with the tm and leaflet packages.

While at first glance it looks in Figure 11.14 like banks are avoiding a minority neighborhood, there's also a big stretch of majority-white areas just to the right that also have no banks. Could zoning, parks, cemeteries, major highways, etc. be affecting the lack of banks in either or both Census tracts? Are the "under-served" areas primarily residential and less densely populated? Is the pattern more about which areas are wealthier, irrespective of race? Or are those areas also affected by proximity to lower-income neighborhoods?

These are all questions to consider when writing about or broadcasting an analysis of this data. Subject-matter expertise is just as critical as the analysis code when telling stories about data.

11.13 Mapping new political boundaries with leaflet

As populations grow, shrink, and shift, political and administrative boundaries change with them. U.S. Congressional districts are redrawn following the decennial Census, or after legal challenges. Local precincts shift along with housing changes. School districts need to be redrawn.

One of the basic tasks in telling stories about new borders is simply showing your audience those new boundary lines, and letting them know where they vote or where their children will go to school. The steps are similar to the bank mapping done earlier, but adding address-search capabilities can make such a Web map more useful. Here's how to do a choropleth map solely with the leaflet package, including address search.

For this project, I'll use new district boundaries for Framingham, Massachusetts, which voted to change from a town to city form of government in 2017. I received the shapefiles from the town's engineering department.

As of this writing, the development version of leaflet was needed for implementing address search. Install the newer-than-CRAN version from GitHub with `pacman::p_load_current_gh('rstudio/leaflet')` – p_load_current_gh() forces an installation even if an older version exists on your system. Then, if you haven't already, also install a package of additional leaflet functions with `pacman::p_load_gh("bhaskarvk/leaflet.extras"")`.

Next, follow along with these 9 steps:

1. Find the folder with geospatial information about precincts. In this example, the FramPrecincts subdirectory of my working directory appears as it is in this book's GitHub project. If your directory structure for the files is different, change your code accordingly.

2. Read the shapefile into a simple features object called pctgeo_sf with

```
pctgeo_sf <- sf::st_read("FramPrecincts/FramPrecincts.shp")
```

```
## Reading layer `FramPrecincts' from data source `D:\Sharon\My Documents Data Drive
## \BookMarkdown\FramPrecincts\FramPrecincts.shp' using driver `ESRI Shapefile'
## Simple feature collection with 18 features and 4 fields
## geometry type:  POLYGON
## dimension:      XY
## bbox:           xmin: 656966.6 ymin: 2918791 xmax: 689247.5 ymax: 2953709
## epsg (SRID):    NA
## proj4string:    +proj=lcc +lat_1=41.71666666666667 +lat_2=42.68333333333333 +lat_0=41
## +lon_0=-71.5 +x_0=199999.9999999999 +y_0=750000 +datum=NAD83 +units=us-ft +no_defs
```

3. This file was prepared with a specific datum (model of the earth's shape) and coordinate reference system (translating 3D points onto a 2D map). What's most important here is that the shapefile needs to have the same projection as the underlying map – the basemap with streets and such. I often like using Esri map tiles for my background maps. A bit of Googling told me that Esri tiles use the WGS 1984 projected coordinate system. So, I need to *re-project* my local shape file to use the same projection as that underlying map. The tmaptools package's `set_projection()` function will do this:

```
pctgeo_sf <- set_projection(pctgeo_sf, projection = "WGS84")
```

4. Run commands such as `str(pctgeo_sf)`, `glimpse(pctgeo_sf)` or `head(pctgeo_sf)`, and you'll see that pctgeo_sf is a simple features object with 18 rows and 4 columns in addition to geometry. PRECINCT has the format "P1", "P2", etc. and came in as a factor; DISTRICT came in as an integer.

I'll add a new column, `District`, to change the DISTRICT integer to a character string. Why? Since doing mathematical operations with district numbers definitely doesn't make sense, these should be characters or factors, not numbers.

```
pctgeo_sf$District <- paste0("District ", as.character(pctgeo_sf$DISTRICT))
```

5. As we saw earlier, the simplest leaflet map creates a leaflet object with `leaflet()` and then adds underlying map tiles using the function `addProviderTiles()`. There's a list of free basemaps available from the leaflet.extras package, at https://github.com/leaflet-extras/leaflet-providers. For a list of *all* available basemaps, run `names(providers)`. You can also use the `addTiles()` command to use default Open Street Map tiles.

If you run this code to start with

```
leaflet(pctgeo_sf) %>%
  addProviderTiles(providers$Esri.WorldStreetMap)
```

all you'll see is a zoomed-out world map. As we did earlier, you can set a particular longitude/latitude point and zoom level with `setView(long, lat, zoomlevel)`. I chose the long/lat for my favorite ice cream shop roughly in the center of town, and experimented with the zoom level until I found one that seemed to show almost all of the town in the map. That zooms the map in to show the town. To add the shapefile's boundary lines, use the `addPolygons()` function.

```
leaflet(pctgeo_sf) %>%
  addTiles() %>%
  setView(-71.4366, 42.3011, 13) %>%
  addPolygons()
```

`addPolygons()` is the critical function here. Without any arguments specifying what those polygons should look like, you'll just see thick blue boundary lines as in Figure 11.15. I want to change that line color and width, and also fill in each district with a different color.

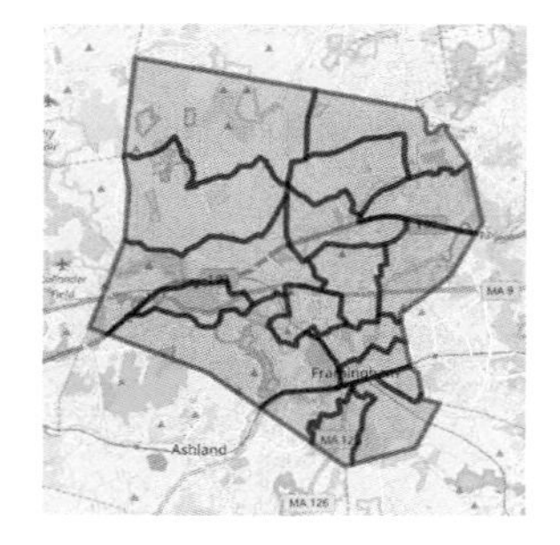

Figure 11.15: Polygons without specifying how to fill them.

6. To change the way the boundary lines look, add arguments to the addPolygons() function. I often start with RStudio's sample map in the leaflet package documentation and then tweak if I feel the need: `addPolygons(color = "#444444", weight = 1, smoothFactor= 0.5, opacity = 1.0)`. Let's take a look at what each of those arguments does.

`color` is for the polygons' *boundary lines*, and not the color inside the polygons. `weight` is for setting line width. `smoothFactor` is how much to simplify those lines (a balance between performance and accuracy, as the documentation explains), and `opacity` is also for the lines.

7. To set the fill color(s) of the polygons themselves, you need to create leaflet *color mapping*. There are a few different ways to do this, depending on the type of data you have.

For a basic district map, you probably just want each district to have a different color – it doesn't really matter which color goes with what district. If you wanted to show percent minority population, income levels, or other *numerical data by district*, the colors would need to correspond to specific numbers or numerical ranges.

These functions tell leaflet how to map colors to data:

`colorFactor()`: For categories that don't necessarily have numerical values that mean anything, such as in this map. These districts are technically numbered, but the fact that one is called "1" and another is called "9" isn't mathematically important.

`colorBin()`: For numerical data that you want to break into a specific number of groups, such as low income, middle class, and high income or $15,000 to $24,999, $25,000 to $49,999, etc.

`colorQuartile()`: Similar to bins but using R's quantile function.

`colorNumeric()`: What you'd expect – mapping numerical data that's continuous, such as median income, but from lowest to highest using a gradual ramp of colors instead of bins.

Once you pick your function, you need two other pieces of information: the color `palette` you want to use (will the map display reds? greens? blues to reds?) and all possible values that might be mappable – in leaflet jargon, the *domain*.

To pick your palette, you can use RColorBrewer or virdis palettes discussed in Chapter 9, select one of R's built-in palettes such as `topo.colors()`, or set some or all of the colors manually. To keep things simple, I'll use the pre-defined `topo.colors()` R palette and will generate 9 colors for my 9 districts. The domain (possible values to map) will be all my districts, the `pctgeo_sf$District` column:

```
district_palette <- colorFactor(topo.colors(9), pctgeo_sf$District)
```

That code creates a palette of colors for all possible values in the data that might be mapped. It doesn't actually match those colors to values on your map.

To *use* this palette in a leaflet map, you need the `addPolygons() fillColor` argument. The syntax is something like `fillColor = ~mypalette(myDataColumnName)`. Note the ~ tilde before the palette name.

For the district map, you can use a command such as `addPolygons(fillColor = ~district_palette(District))`.

By now you may understand why I like the tmap package – I find its syntax simpler and more straightforward. However, leaflet is so powerful and customizable, it can be worthwhile to learn if you want to do robust mapping with R. And, since this is scripting, you just need to do a map like this *once*. Then it should be fairly easy to swap in new data and make small tweaks. *I can't emphasize this enough:* It can really help your R efficiency either to make a code snippet or to save your template files in a place where they're easy to find when you need them again.

The map code now so far:

```
district_palette <- colorFactor(topo.colors(9), pctgeo_sf$District)

leaflet(pctgeo_sf) %>%
  addTiles() %>%
  setView(-71.4366, 42.3011, 13) %>%
  addPolygons(color = "#444444", weight = 1, smoothFactor= 0.5, opacity = 1.0,
   fillColor = ~district_palette(District)
  )
```

8. Now the districts are mapped. But since this is a Web map, it would be nice if viewers could find out which district a region is in by simply clicking or mousing over an area, instead of having to look at a color-coded legend. For labels that appear when the user mouses over (or taps on a mobile device), use `addPolygons()`'s `labels` and `labelsOptions()`.

`labels` sets up the text that you want to appear. I'd like both the District and Precinct information to be included in the information box. Currently, precincts are in the PRECINCT column as factors such as "P1", "P10", and so on. I'd like to turn the factors into character strings with as.character(), and then get rid of the initial "P" in each string.

There are several easy ways to eliminate the P, including deleting the first character or searching for P and replacing it with nothing. I'll delete the "P" character with base R's gsub() (global substitute) function.

gsub() uses the syntax `gsub("SearchString", "ReplacementString", "String to operate on")`.

```
pctgeo_sf$Precinct <- gsub("P", "", pctgeo_sf$PRECINCT)
```

Leaflet labels use HTML. Looking at some examples, I see that this means I'll want to run the HTML function from the htmltools package on the text. This lets leaflet know that a character string should be treated as HTML. In addition, once I set up all the label text, I can use lapply() to run the HTML function on each one of the labels.

As long as I'm using HTML, I can do a little styling. I'd like the labels to have District in bold on the first line and Precinct in plain text on the next line. So, the format would be:

```
DistrictLabels <- glue::glue("<b>{pctgeo_sf$District}</b><br />Precinct {pctgeo_sf$Precinct}") %>%
  lapply(htmltools::HTML)
```

This created label text. Note that I used the glue package to make it easy to combine text strings and variables into a single string, and then applied the HTML function with base R's lapply.

To *use* your labels in the map, add them with the label argument within addPolygons(). Here's the map code for a map like in Figure 11.16:

```
leaflet(pctgeo_sf) %>%
  addTiles() %>%
  setView(-71.4366, 42.3011, 13) %>%
  addPolygons(color = "#444444", weight = 1, smoothFactor= 0.5, opacity = 1.0,
   fillColor = ~district_palette(District),
   label = DistrictLabels
  )
```

There's a lot more formatting and styling you can do for the labels info box with the `labelOptions` argument. RStudio's sample, for example, uses the code below to set things like text size and font weight.

```
labelOptions = labelOptions(
    style = list("font-weight" = "normal", padding = "3px 8px"),
    textsize = "15px")
```

If you don't want the label to appear unless the user clicks an area, use `popup = DistrictLabels` instead of `label = DistrictLabel`.

9. It might be helpful for users to be able to enter an address into the map, instead of having to hunt around to find their homes or other places of interest. The leaflet.extras package offers a lot of added functionality to leaflet mapping in R, using plug-ins written for the JavaScript leaflet.js library. Among the add-ins: three different types of address searching.

The easiest to implement uses OpenStreetMap. Because of the "open" nature of that project, you don't need an API key or to authenticate an account. However, it generally only zooms to a city or street, but not a specific address. That's useful if, say, you've got a national map color-coded by county; but not helpful for a local precinct map where a street can be in more than one precinct.

Package developer Bhaskar Karambelkar also added Google and Bing searching to leaflet.extras. To add a Google Maps address search bar to your leaflet map, you'll first need to sign up for a Google Maps API key. A standard, free account allows 25,000 uses per day, which might not be enough for a popular map at a high-traffic Web site, but should be plenty for a smaller community or personal site. Google has full instructions on how to sign up for an API key at https://developers.google.com/maps/documentation/javascript/get-api-key. That will also tell you how to enable the key for a project in the Google Developer Console.

Once you've done that, the rest is pretty easy. leaflet.extra's Google Maps search function will be looking for your API key in an *environment variable* called `GOOGLE_MAP_GEOCODING_KEY`. Without going into too much technical detail about the difference between a "regular" R variable that we've been using so far and this new type of environmental variable, I'll just say that the variables we've been using up until now in RStudio's

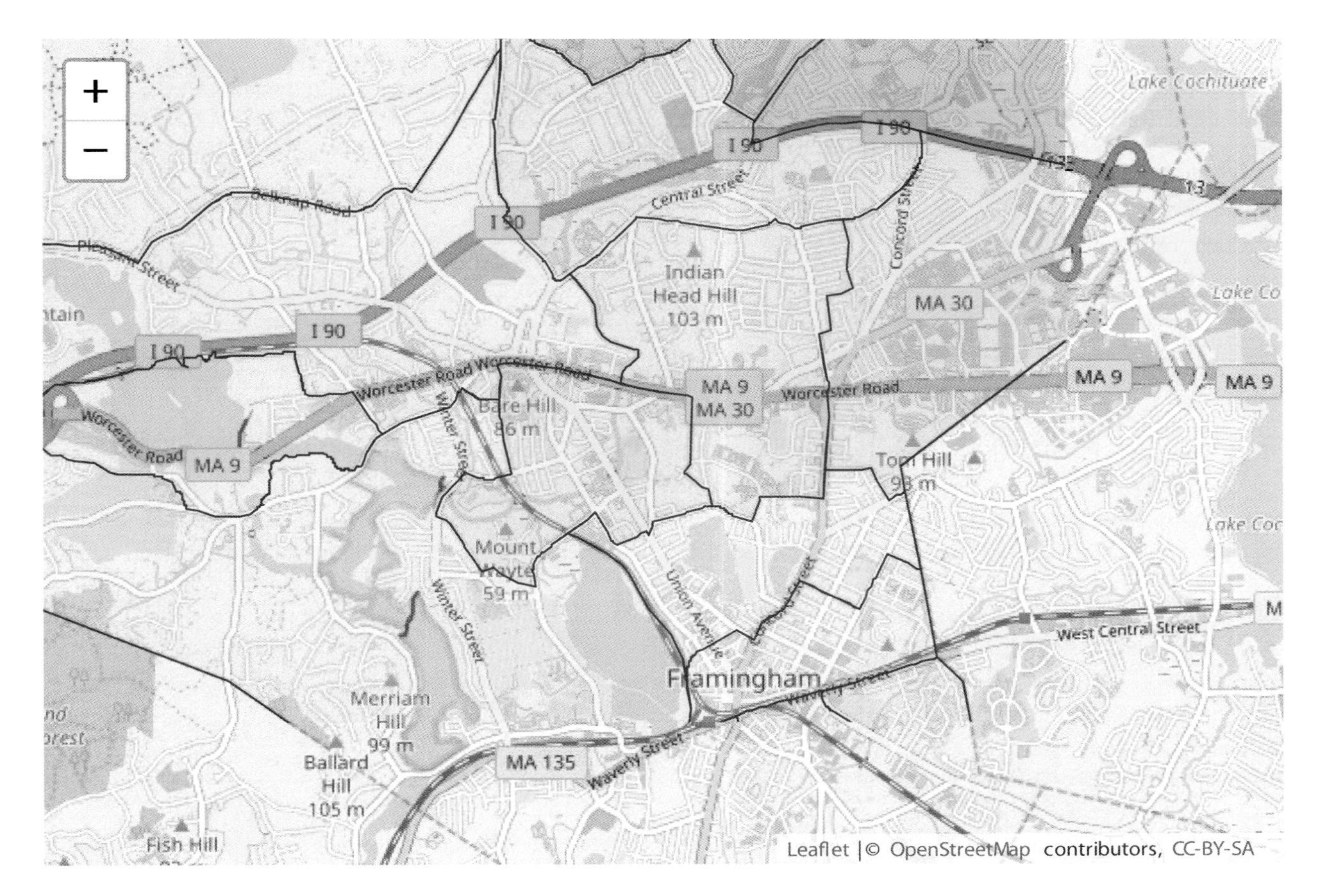

Figure 11.16: leaflet map of precincts.

interactive console have lived in what's called the "global environment." To set this new type of variable, you use base R's `Sys.setenv()` function:

```
Sys.setenv(GOOGLE_MAP_GEOCODING_KEY = "YourGoogleAPIKey")
```

Or, you can use `usethis::edit_r_environ()` to open the environment file and add `GOOGLE_MAP_GEOCODING_KEY = "YourGoogleAPIKey"` to the file.

Hadley Wickham's Advanced R book (part of Chapman & Hall's R Series) has an entire chapter on environments if you'd like to learn the ins and outs of environments in R. For this purpose, though, you can just run the code above.

You can also add Sys.setenv() code to a special *.Rprofile* file in your working directory. R will look for this file whenever you're starting to work in this particular project directory. Or, you can use a global Rprofile.site file (see the startup helpfile with ?Rprofile.site).

Note: There are lots of things you can customize in an Rprofile start-up. I don't recommend doing things like auto-loading packages you often use, because if you depend on that in your scripts, those scripts won't work on other machines without that Rprofile file.

Rprofile is useful for other things, though. For example, you can customize the width of your console with a line in Rprofile.site: `options(width = 120)`. This makes the console 120 characters wide instead of the 80-character default, and can make some data printout more legible. And that won't affect the usability of your R code on other machines.

Once your API key value is set, you can load leaflet.extras and add `%>% addSearchGoogle()`

at the end of your map. Or, if you'd prefer not to load leaflet.extras into memory, use `%>% leaflet.extras::addSearchGoogle()` to the map. Another command, `addResetMapButton()`, creates a button to zoom the map back to its original state.

The full leaflet code, assuming your Google API key is properly stored:

```
leaflet(pctgeo_sf) %>%
  addTiles() %>%
  setView(-71.4366, 42.3011, 13) %>%
  addPolygons(color = "#444444", weight = 1, smoothFactor= 0.5, opacity = 1.0,
   fillColor = ~district_palette(District),
   label = DistrictLabels
  ) %>%
  addSearchGoogle() %>%
  addResetMapButton()
```

If you'd like to add point markers for polling places, or schools for new school districts, you'd use leaflet's addMarkers() function, as described in the bank map.

This map can also be saved to a stand-alone HTML file with the htmlwidget package's saveWidget() function:

```
framingham_map <- leaflet(pctgeo_sf) %>%
  addTiles() %>%
  setView(-71.4366, 42.3011, 13) %>%
  addPolygons(color = "#444444", weight = 1, smoothFactor= 0.5, opacity = 1.0,
   fillColor = ~district_palette(District),
   label = DistrictLabels
  ) %>%
  addSearchGoogle() %>%
  addResetMapButton()

saveWidget(framingham_map, "framingham_new_precinct_map.html")
```

You could then embed the map in any HTML page with an iframe.

11.14 Inspiration: Washington Post investigation

The first project added to the Washington Post's investigative reporting unit's GitHub repository was done in R: a look at how Jared Kushner and his partners used a program meant for high-unemployment areas to build a luxury skyscraper in Jersey City, N.J. You can see the mapping and census data analysis at https://github.com/wpinvestigative/kushner_eb5_census.

11.15 Wrap-up

We covered the tmap, tmaptools, and leaflet packages for making both static and interactive maps. That included importing custom geographies and geocoding addresses, as well as importing U.S. Census Bureau geospatial files with tigris. We also learned R's paste() and paste0() functions for combining text, as well as how to geocode addresses within R.

Next up: Election data in R

11.16 Additional resources

For more examples of using the tidycensus package for U.S. Census data, try Julia Silge's post, Using tidycensus and leaflet to map Census data, https://juliasilge.com/blog/using-tidycensus/.

And for more on using tmap, I wrote a tutorial for Computerworld at https://www.computerworld.com/article/3038270/data-analytics/create-maps-in-r-in-10-fairly-easy-steps.html. Zev Ross posted one at http://zevross.com/blog/2018/10/02/creating-beautiful-demographic-maps-in-r-with-the-tidycensus-and-tmap-packages/. It includes the helpful tip that tm_shape() can re-project your data directly with the projection argument, such as tm_shape(my_sf, projection = "WGS84"). And, Thomas Lo Russo posted a nice how-to from a workshop he did with Max Grutter, https://tlorusso.github.io/geodata_workshop/tmap_package.

There are several other excellent mapping packages in R that you might want to investigate (I didn't have space to demo them all):

ggplot2 added support for simple features in the CRAN version in 2018. For exploratory analysis, creating a choropleth map is as simple as `ggplot(simpleFeaturesObject) + geom_sf(aes(fill = columnName))`.

The **choroplethr** package, as the name implies, is designed to create choropleth maps. It was designed specifically to map U.S. Census data, but now handles importing shapefiles as well. http://www.arilamstein.com/open-source/. Creator Ari Lamstein demos animated choropleths at http://www.arilamstein.com/open-source/choroplethr/animated-choropleths/.

The **plotly** general data visualization R package also does mapping. You can see more info at https://plot.ly/r/choropleth-maps/. Package maintainer Carson Sievert blogged about using plotly for mapping with ggplot2 at https://blog.cpsievert.me/2018/01/30/learning-improving-ggplotly-geom-sf/ and with simple features at https://blog.cpsievert.me/2018/03/30/visualizing-geo-spatial-data-with-sf-and-plotly/.

ggmap For more on using ggmap with ggplot2, see the GitHub repo at https://github.com/dkahle/ggmap.

For an excellent example of geospatial analysis and housing prices, I'd suggest the Urban Spatial blog's detailed examination of San Francisco Housing prices. It was done by Ken Steif, director of the Master of Urban Spatial Analysis program at the University of Pennsylvania, and then-Masters student Simon Kassel did a sophisticated analysis of San Francisco housing prices by neighborhood in R. http://urbanspatialanalysis.com/dataviz-tutorial-mapping-san-francisco-home-prices-using-r/

Len Kiefer, Deputy Chief Economist at Freddie Mac, does a lot with mapping economic and housing data on his blog, such as the collection at http://lenkiefer.com/2016/08/24/more-maps.

For technical details on mapping in general as well as in R:

- 2-page primer on **projections and datums** from University of California, Berkeley's Geospatial Innovation Facility: http://gif.berkeley.edu/documents/Projections_Datums.pdf.
- Melanie Fraier's 4-page overview of **Coordinate Reference Systems in R**, which includes datums and projections: https://www.nceas.ucsb.edu/~frazier/RSpatialGuides/OverviewCoordinateReferenceSystems.pdf. Frazier is a scientific programmer at the National Center for Ecological Analysis and Synthesis at University of California, Santa Barbara. She also has a ggmap Cheatsheet at https://www.nceas.ucsb.edu/~frazier/RSpatialGuides/ggmap/ggmapCheatsheet.pdf and an R Meetup PowerPoint presentation on ggmap for download at https://www.nceas.ucsb.edu/~frazier/RSpatialGuides/ggmap/QuickPointMapping.pptx.
- For an entire book devoted to geospatial analysis in R, check out Geocomputation with R by Robin Lovelace, Jakub Nowosad, and Jannes Muenchow, due out from CRC Press. The book's Web site is at https://geocompr.robinlovelace.net/.
- And finally, if you want a thorough technical understanding of mapping in R and the sf package, watch the two-part video tutorial "Spatial data in R: new directions" by Edzer Pebesma, lead author of the sf package, from the 2017 useR! conference. You should be able to find it on the Microsoft Channel 9 video site by searching at https://bit.ly/RMappingVideo.

Chapter 12

Putting it all Together: R on Election Day

Data is one of the highlights of Election Day coverage, whether you work at a newspaper, radio station, TV, or community blog. Many others such as political staffers and community activists are eager to analyze this data as well, making it a great topic for taking R skills covered so far for a test drive.

Even if you don't work with political data, though, you'll likely find skills in this chapter helpful for many other types of analysis.

12.1 Project: Election data

The goal of this chapter is to come up with a template, or recipe, for how you might want to structure an election analysis. Once you've got the recipe, it should be easy to modify for covering other, similar races.

12.2 What we'll cover

- Planning for Election Day
- Creating an R script "recipe" for analysis
- Using base R's save() and load() functions
- Generating interactive tables with a single line of code
- Calculating and visualization correlations
- Creating quick interactive graphs with plotly and highcharter

12.3 Packages needed in this chapter

```
pacman::p_load(dplyr, ggplot2, magrittr, tmap, tmaptools, leaflet, stringr, janitor,
               readr, DT, rio, htmlwidgets, plotly)
pacman::p_load_gh("smach/rmiscutils")
pacman::p_load_gh("hrbrmstr/taucharts")
```

12.4 Election Day preparation

With some pre-election planning, R can help you generate on-deadline charts, graphs, and analysis. This goes the same for other expected data you'd like to analyze on a tight time frame, such as new census population data or the latest job-growth report. Planning is critical.

Sweden's Aftonbladet newspaper went all-in with R for its election coverage back in 2014. In a blog post about their experience, Jens Finnäs said a key to successful on-deadline coverage was preparation. They decided to focus on breakdowns by municipality, and collected some data in advance. They also decided what questions they wanted to answer *before* results started coming in.

Additional ways you can prepare in advance:

In what format will election results be available? Knowing this will give you a head start in creating charts and graphs. Will data be in the right shape for you to analyze in your preferred R tools? If not, do you have code ready to get it prepped for analysis? If you're pulling in national/state-wide election results from a paid service API or compiling results yourself from staff reports in the field, you may know the format in advance. If results come from a local official, maybe the format is similar each year; or you can ask in advance for a sample spreadsheet.

What would you like to do with this data? Create a searchable, sortable chart? Graphs? Maps? For election data, do you want to examine how demographics affect candidate preferences? Show which communities changed party preferences over time? Look at turnout versus demographics or results? Using results from a prior year or test data, you can set up code ahead of time to do many of these things and then swap in updated results.

This chapter's project shows how I might have prepared to cover the 2018 Massachusetts gubernatorial race. I'll use 2014 data to visualize results and create an interactive table. I'll then have steps that will be fairly easy to adapt for upcoming contests.

12.4.1 Step 1: Configure data files

When cooking a meal, things work out better if you have all the necessary tools and ingredients at the start, ready on the counter as you go through your steps. It's the same with an R "recipe" – I like to set up my data variables at the outset, making sure I've got "ingredients" at hand.

For my recipes, I start by creating variables with data files at the top, making it easier to re-use with other, similar data (as well as move my work from one computer to another). Also at the top, I load all packages I want in memory.

Where's your data? I suggest giving some thought to how you want to store your data files. I spent a few years creating one R project for each election I analyzed, ending up with data and geospatial files scattered all over my computer's hard drive. That made it unnecessarily aggravating to find geospatial files that I wanted to reuse, or pull together past election results over several years.

Later, I made one folder in my Documents directory for geofiles and another for election data, each with subdirectories for national, state, and local files. Files I'd like to use on both my desktop and laptop are in a Dropbox folder that syncs on both systems.

This may not work for you. You might want to create one R project for mapping files and another for election data. When you get more advanced (beyond what I'll cover in this book), one of the best things you can do is create your own R package. However you decide to organize your data, it's helpful to keep to a scheme so you know where your data and scripts are if you want to use them again.

Get the data. Data from past Massachusetts gubernatorial elections are available from the Secretary of State's office with results by county, municipality, and precinct. I downloaded the data and included them as Excel files in this book's GitHub repo.

I added this to the top of my *script* file:

```
##### CONFIGURATION #####

#### Voting results file locations ####

ma_gov_latest_available_file_name <- "data/PD43+__2014_Governor_General_Election.csv"
ma_gov_previous_available_file_name <- "data/PD43+__2010_Governor_General_Election.csv"
```

Next, I'd like a geospatial file of Massachusetts cities and towns. You can import that data into R with the tigris package. Somewhat non-intuitively, getting boudaries for Massachusetts' 351 cities and towns requires tigris's `county_subdivisions()` function:

```
ma_cities <- tigris::county_subdivisions("MA", cb = TRUE, class = "sf")
```

I'm pretty sure I'll want to use these again someday. I can save the information and R structure using base R's save() function. `save(ma_cities, file = "path/to/geofiles/MA/mamap2015.Rda")` saves the ma_cities object to a file in a project subdirectory called `mamap2015.Rda`. Multiple objects can be stored in a single file using the syntax `save(variable1, variable2, variable3, file = "path/to/myfile.Rda")`. You can also use the longer `.Rdata` file extension instead of `.Rda` if you'd like.

When you want to use an .Rda (or .Rdata) data object again that you created with the `save()` function, run `load("path/to/geofiles/MA/mamap2015.Rda")` (or wherever you saved it). Any R objects stored in mamap2015.Rda will be loaded into your working session, using the variable names you used to store them.

I'll add this to my script:

```
#### Map file locations ####
ma_geospatial_object <- "path/to/geofiles/MA/mamap2015.Rda"
```

There are a couple of other things you might want to set up as variables in this configuration section if you're serious about re-use. Adding information such as chart and map headlines and source information makes it easier to re-use the script in a future year with new data from the same sources.

```
#### Headline and source info ####
ma_gov_headline <- "Massachusetts 2014 Governor's Results"
ma_gov_datasource <- "Source: Massachusetts Secretary of State's office"

##### END CONFIGURATION #####
```

12.4.2 Step 2: Load packages

I usually want dplyr and ggplot2 when working with data, so I tend to load them in most R scripts I write. If you've got favorite R packages, you might want to do the same. For this project, I'd like the option of using magrittr for its elegant piping syntax, which I'll cover in Step 4. And I expect to use tmap, tmaptools, and leaflet for mapping. I'll add more packages to this section as I go on, but I'll begin by adding this to my recipe script:

```
#### Load packages ####

library(dplyr)
library(ggplot2)
library(magrittr)
library(tmap)
library(tmaptools)
library(leaflet)
library(stringr)
```

```
library(janitor)
library(readr)
library(DT)
```

Reminder: Using four or more pound signs *after* each section header lets me use navigation at the bottom of the script panel, as I explained in Chapter 6.

12.4.3 Step 3: Import data.

I imported both 2010 and 2014 voting results into R using the *variables* that hold the file name and location: `ma_gov_latest_available_file_name` and `ma_gov_previous_available_file_name`.

```
##### IMPORT DATA #####

#### Import election results ####
ma_latest_election <- rio::import(ma_gov_latest_available_file_name)
ma_previous_election <- rio::import(ma_gov_previous_available_file_name)
```

When I've got the latest 2018 results, I can swap that file into my `ma_gov_latest_available_file_name` variable at the top of the script, and I won't have to worry about remembering to change it anywhere else.

I'll also import the shapefile here with with

```
#### Import Massachusetts cities geospatial object ####
load(ma_geospatial_object)
```

12.4.3.1 Syntax shortcut

The pattern of taking a data frame, performing some operation on it, and storing the results back in the original data frame is such a common one that there's a piping shortcut for it. `mydf %<>% myfunction(argument1, argument2)` is the same as

```
mydf <- mydf %>%
  myfunction(argument1, argument2)
```

The `%>%` operator is available for use whenever you load the dplyr package. The `%<>%` operator isn't. To use `%<>%`, you need to explicitly load the magrittr package with `library(magrittr)`.

This syntax may make for less readable code, especially when you're first starting out. However, it also makes for less typing – and that means less possibility for errors, especially if you have multiple objects with similar names. Decide which syntax works best for you for the project at hand.

12.4.4 Step 4: Examine (and wrangle) the data

I'll start checking the imported data with `str(ma_latest_election)`, `head(ma_latest_election)`, `tail(ma_latest_election)`, and `summary(ma_latest_election)` (or `glimpse(ma_latest_election)` or `skimr::skim(ma_latest_election)`). I'll just work with the 2014 data in this section to save space; the 2010 data would be similar.

When I ran str(), head(), and tail(), I saw a few issues. There are 352 rows instead of 351, and – from tail() as you can see in Figure 12.1 – the last row is a total row that I don't want for the map (that explains why there are 352 rows when I should only have results for 351 cities and towns). Most importantly, though, all the results columns came in as character strings instead of numbers.

```
tail(ma_latest_election)
```

```
##        City/Town V2 V3 Baker/ Polito Coakley/ Kerrigan Falchuk/ Jennings
## 347       Woburn NA NA         7,381             5,772                383
## 348    Worcester NA NA        16,091            20,297              1,309
## 349 Worthington NA NA           228               310                 38
## 350     Wrentham NA NA         3,037             1,325                139
## 351     Yarmouth NA NA         5,426             4,164                237
## 352       TOTALS NA NA     1,044,573         1,004,408             71,814
##     Lively/ Saunders Mccormick/ Post All Others Blank Votes
## 347              101             105         11         145
## 348              353             345         57         527
## 349                4               1          0          14
## 350               34              39          5          39
## 351               53             212          7          97
## 352           19,378          16,295      1,858      28,463
##     Total Votes Cast
## 347           13,898
## 348           38,979
## 349              595
## 350            4,618
## 351           10,196
## 352        2,186,789
```

Figure 12.1: Viewing the last few rows of election data with tail().

This is one of the few cases where Excel is "smarter" than an R package. Open up the CSV in Excel, and it understands that "25,384" is a number. R, however, expects numbers to only have digits and decimal places. Once it sees a comma, it will "assume" a string unless there's some code to tell it otherwise.

Fortunately, there's an R package that understands numbers with commas by default when importing data: readr. So let's re-import the files using readr's read_csv() function:

```
ma_latest_election <- readr::read_csv(ma_gov_latest_available_file_name)
ma_previous_election <- readr::read_csv(ma_gov_previous_available_file_name)
```

The total row is easy to get rid of. I could remove row 352 with `ma_latest_election <- ma_latest_election[-352,]` using bracket notation or `slice(ma_latest_election, -352)` with dplyr. Or, as in the last chapter, I could slice off the final row using dplyr's n() function and the minus sign: ma_latest_election <- slice(ma_latest_election, -n()).

But to make a general recipe, I could also filter out any row that starts with "total", "Total", or "TOTAL".

I'm not sure whether a total row will have "total", "Total", or "TOTAL". But I don't have to worry about that if I explicitly tell R to ignore case. The stringr package's str_detect() function comes in handy here, with the syntax `str_detect("mystring", "mypattern")` when using case-sensitive matching and `str_detect("mystring", regex("mypattern", ignore_case = TRUE))` to ignore case.

I'm also using the `%<>%` operator here.

```
ma_latest_election %<>%
  filter(!str_detect(`City/Town`, regex("total", ignore_case = TRUE)))
```

If that code seems a little complicated, another way to do this is to temporarily change the City/Town column to all lower case and then use the simpler str_detect syntax:

```
ma_latest_election %<>%
  filter(!str_detect(tolower(`City/Town`), "total"))
```

tolower() here doesn't permanently change the column to all lower case, but transforms it only for the purpose of checking for string matching. `!` before str_detect signifies "doesn't match".

Next: It *looks* like there's no data in columns X2 and X3. You can check to make sure by running `unique(ma_latest_election$X2))` and `unique(ma_latest_election$X3))` to view all the unique items in that column, or `table()` on each column to see a table of all available data.

To delete a column, you can manually set its value to NULL:

```
ma_latest_election$X2 <- NULL
ma_latest_election$X3 <- NULL
```

For deleting more than one column, dplyr's select() might be a better choice: `ma_latest_election <- select(ma_latest_election, -X2, -X3)`. The minus sign before column names means they should be *removed* instead of selected.

Even better, though, would be a function that deletes all columns that have nothing in them. There's a function in the janitor package that does this: remove_empty().

```
ma_latest_election <- remove_empty(ma_latest_election, which = "cols")
```

As you might have guessed, `remove_empty(ma_latest_election, which = "rows")` would remove all empty rows.

R-friendly column names. To get ready for my main analysis, I'll rename the columns to make them R friendly (getting rid of slashes and spaces) as well as chart friendly.

```
names(ma_latest_election) <- c("Place", "Baker", "Coakley", "Falchuk", "Lively",
           "Mccormick",  "Others", "Blanks", "Total")
```

If I'm re-using this script in future elections, I'll need to make sure to replace this with new candidate names. One strategy is to put the column names in the configuration area at the top of the file. Another is to mark every area that needs updating with a comment such as `#### NEEDS REPLACING ####`.

12.4.5 Step 5: Who won and by how much?

The previous section got the data ready for analysis in R. This section will do a few simple calculations to make it easy to see who won and by how much.

These are things I need to do with every spreadsheet of election results, unless that spreadsheet already includes columns for winners, vote percentages, and margins of victories. Instead of re-writing that code each time I analyze election data, I created a couple of functions in an R election utilities package that I can re-use. With those ready-to-use functions for processing election data, I find election calculations in R even easier than doing them in Excel.

I wrote these functions years ago, before creation of packages like purrr. The code is old, and could probably be rewritten more efficiently, but it works. You are welcome to use those functions, too, if you'd find them helpful, which are in my rmiscutils package.

Calculating a winner per row. It's a multi-step operation in R to look across each row, find the largest vote total, get the name of the column with the largest number, and make sure to account for ties before determining who won in each row. The rmiscutils package's elec_find_winner() function does this for you.

elec_find_winner() takes these arguments: 1) *name* of a file *as a character string* (in quotation marks) because it assumes you're using the package to read in a CSV or Excel file, although `"mydata"` without a .csv or .xlsx extension will assume you want to use an already-existing data frame and convert the character string into an R data frame object for you; 2) the *number* of the column where candidate vote totals start; 3) the *number* of the column where candidate vote totals end; and 4) whether you want to export the results to a CSV file.

This function calculates a winner for each row, as long as the data is formatted as the function expects: each candidate's vote total is in its own column, and each place (city, precinct, state, etc.) is in its row. And, it returns both candidates' names if there's a tie.

For the 2014 Massachusetts election data, the command would be:

```
winners <- elec_find_winner("ma_latest_election", 2, 5, FALSE)
```

By specifying columns 2:5, I excluded write-ins, blanks, and the Total column when calculating the winner. The function creates a new Winner column.

I can do a quick check of how many cities and towns each candidate won by counting items in the Winner column with base R's table() function:

```
table(winners$Winner)
```

```
##
##   Baker Coakley
##     232     119
```

Victory margins. This data has raw vote totals, but I'm also interested in vote percentages for each candidate in each place. Percents can answer the question **"Where were each candidate's strengths?"**

Percents can also better show at a glance how polarized an outcome was. A race can be close overall *and* close in most communities, or it can be close overall but with many communities voting heavily for one side or the other. Looking at a map or chart showing only which candidate won misses a key part of the story: Winning by 1 point isn't the same as winning by 12.

Other interesting questions that percent margins-of-victory can answer, especially for a close election: **"How close was *each community*?"**" (or state, or precinct, depending on your level of analysis), and **"How many communities were competitive?"**

But percents alone miss another part of the story, because there's a difference between winning by 5 points in Boston and winning by 5 points in Bolton. Electoral strength in a major city can be worth tens of thousands of votes in padding a victory margin; winning a small town by the same margin could net just a few dozen. So, I also like to calculate how many *net votes* the winner received versus his or her nearest challenger in each community. That answers the question, **"What areas helped most in winning the election?"**

In Massachusetts, Boston has the largest number of votes of any community by far, and Democrats almost always win there. But for Democrat Martha Coakley in 2014, winning Boston handily didn't give her an overall victory – she didn't *run up her overall vote total enough* to offset Republican Charlie Baker's strength elsewhere. Comparing results with prior elections would be helpful here to understand what happened.

One possibility is that she won Boston by a slimmer percent margin than Democratic Gov. Deval Patrick did before her. Another possibility is that she won Boston by similar margins as Patrick did, but turnout was lower in the city than in prior years. Yet another possibility is that all was the same in Boston as when Patrick won, but pro-Baker areas had higher turnout.

However, one step at a time. Let's first get 2014 data in shape for some basic analysis.

My rmiscutils package's elec_pcts_by_row() function will generate percents for each value across a row. If the data is formatted like this

```
##       Place Baker Coakley Total
## 1 Abington  3459    2105  5913
## 2     Acton  3776    4534  8736
## 3 Acushnet  1500    1383  3145
```

elec_pcts_by_row() can calculate Baker's percent and Coakley's percent versus the overall Total in each row. The syntax is `elec_pcts_by_row(mydata, c("candidate1", "candidate2"), "TotalColumn")` (the total column name expects the total column to be named "Total" so it won't be necessary to define it here.) Note

I'm also using magrittr's `%<>%` so I don't have to keep writing out `winners <- winners %>%` , but if that's less understandable for you, feel free to keep using "wordier" syntax.

```
library(dplyr)
library(magrittr)
winners <- elec_pcts_by_row(winners, c("Baker", "Coakley"))
winners %<>%
  mutate(
    Baker.pct.margin = Baker.pct - Coakley.pct,
    Baker.vote.margin = Baker - Coakley
  )
```

I'm not interested in the third-party candidates, write-ins, blanks, or the lower-case place column right now, so I'll remove columns 4 through 8 for streamlined analysis.

If you want to use this for analyzing another election, you could simply replace "Baker" and "Coakley" here; and then change the new column names from Baker.pct.margin and Baker.vote.margin to Winner.pct.margin and Winner.vote.margin.)

```
winners<- winners[,-c(4:8)]
```

A basic summary() will check the data (see Figure 12.2):

```
summary(winners)
```

```
##      Place                Baker          Coakley               Total
##  Length:351          Min.   :   20   Min.   :     11.0   Min.   :    36
##  Class :character    1st Qu.:  704   1st Qu.:    691.5   1st Qu.:  1654
##  Mode  :character    Median : 2189   Median :   1424.0   Median :  3869
##                      Mean   : 2976   Mean   :   2861.6   Mean   :  6230
##                      3rd Qu.: 4205   3rd Qu.:   3215.0   3rd Qu.:  8286
##                      Max.   :47653   Max.   :104995.0   Max.   :161115
##      Winner            Baker.pct       Coakley.pct     Baker.pct.margin
##  Length:351          Min.   :13.30   Min.   :21.10   Min.   :-68.800
##  Class :character    1st Qu.:38.55   1st Qu.:34.00   1st Qu.:-12.850
##  Mode  :character    Median :53.70   Median :39.90   Median : 13.800
##                      Mean   :49.22   Mean   :43.75   Mean   :  5.472
##                      3rd Qu.:59.50   3rd Qu.:51.80   3rd Qu.: 25.250
##                      Max.   :69.80   Max.   :82.10   Max.   : 48.200
##  Baker.vote.margin
##  Min.   :-57342.0
##  1st Qu.:  -105.5
##  Median :   431.0
##  Mean   :   114.4
##  3rd Qu.:  1227.0
##  Max.   :  4222.0
```

Figure 12.2: A summary of the winners data frame.

Next, there are some other dataviz "basics" I'd like to run on election data to find out who had the largest vote margins and percent victory margins, and where it was closest.

12.4.6 Step 6: Exploratory visualizations

There are two different types of visualizing you're probably going to want to do if you're covering an election. *Exploratory* visualizations will help you better understand the data and decide what stories you want to tell with it. These don't have to be pretty or well labeled; they just need to help you see patterns. *Presentation* visualizations are for your audience, and so need to be constructed with a bit more care.

For quick visualizations, base R's hist() on the Baker.pct.margin column will give a sense of how many communities were close vs. blowouts:

```
hist(winners$Baker.pct.margin)
```

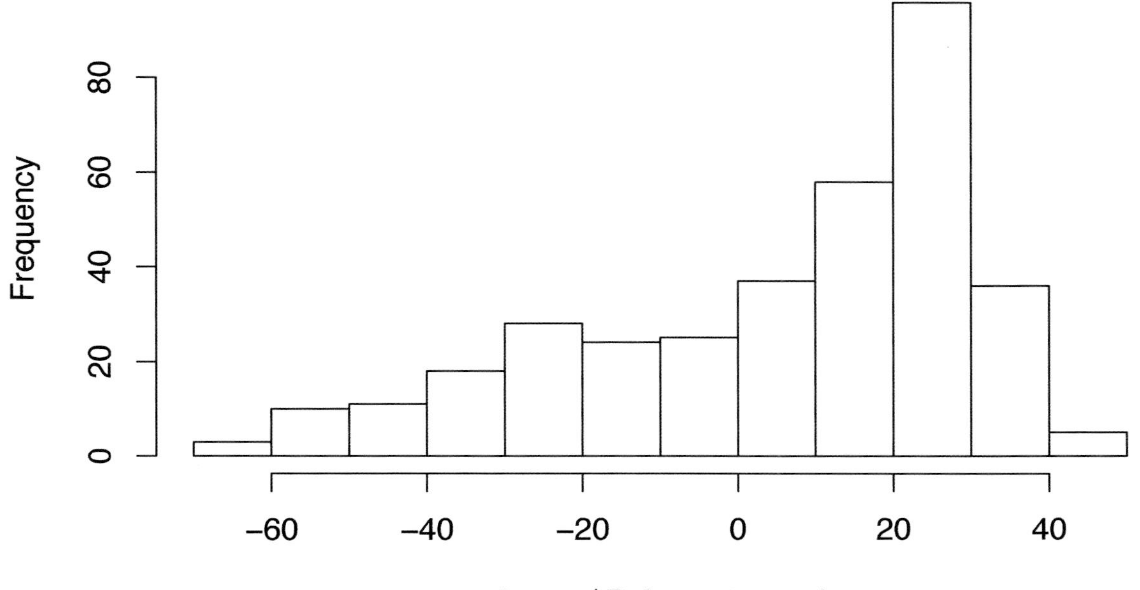

Figure 12.3: A histogram of Baker's winning percent margins.

You can see in Figure 12.3 that Baker had a few big losses/Coakley had a few big wins, but the most common results had Baker winning by between 10 and 30 percentage points.

What about actual margin of victory in raw votes? That's the Baker.vote.margin column:

```
hist(winners$Baker.vote.margin)
```

It looks in Figure 12.4 like Baker won a lot communities by small vote totals – either because large communities were close or he won in a lot of small towns. Later in this chapter, we'll take a look at that.

The way these election results are structured, Baker's largest margins are positive numbers and Coakley's largest margins are negative numbers, because I calculated data for the winning candidate, Charlie Baker. However, I'm also curious to see the largest margins *regardless of who won.* For this case, then, I want the *absolute values* of Baker's vote margins, turning everything into a positive number. Otherwise, if I pull "largest vote margins," I'll only get Baker's top margins and won't see Coakley's top raw-vote wins.

I'll use dplyr's top_n() function to view the top 5 rows. top_n() takes the syntax `top_n(dataframe, numrows, sortingcolumn).` :

```
  top_n(winners, 5, abs(Baker.vote.margin))
```

```
##         Place Baker Coakley  Total  Winner Baker.pct Coakley.pct
## 1      Boston 47653  104995 161115 Coakley      29.6        65.2
```

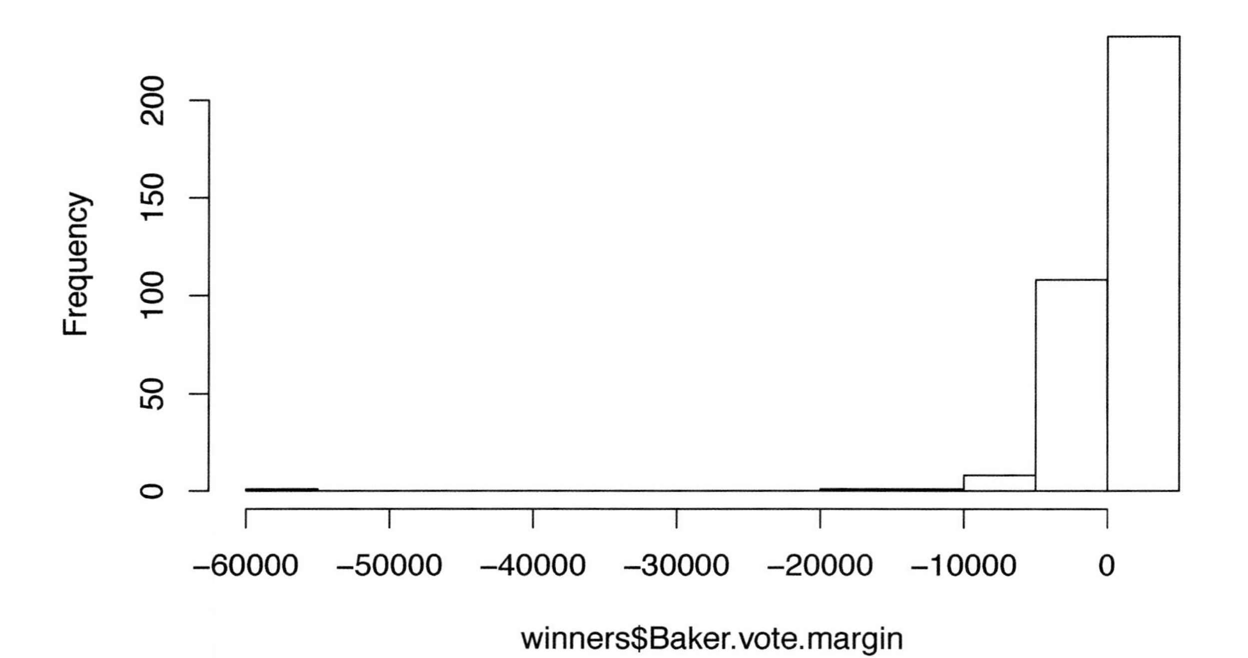

Figure 12.4: A histogram of Baker's winning vote totals.

```
## 2   Cambridge  5589   25525  32725 Coakley       17.1         78.0
## 3      Newton 12089   19068  33019 Coakley       36.6         57.7
## 4  Somerville  4918   16351  22844 Coakley       21.5         71.6
## 5 Springfield 10256   19312  34375 Coakley       29.8         56.2
##   Baker.pct.margin Baker.vote.margin
## 1            -35.6            -57342
## 2            -60.9            -19936
## 3            -21.1             -6979
## 4            -50.1            -11433
## 5            -26.4             -9056
```

Coakley, the loser, had the five largest raw-vote wins, including winning the state's largest city by a very healthy margin.

Unfortunatey, top_n() isn't arranging these from largest to smallest, but you can do that by adding `arrange()` to sort by Baker's vote margin in descending order (don't forget to take the absolute value):

```
top_n(winners, 5, abs(Baker.vote.margin)) %>%
arrange(desc(abs(Baker.vote.margin)))
```

Which communities had the narrowest margins by percent? I'll again take an absolute value, this time of the Baker.pct.margin column, and look at the *bottom* 5 to get the smallest margin. To get the lowest 5 instead of largest 5, use `-5` as top_n()'s second argument and not 5:

```
top_n(winners, -5, abs(Baker.pct.margin)) %>%
  select(Place, Baker.pct, Coakley.pct, Baker.pct.margin) %>%
  arrange(abs(Baker.pct.margin))
```

```
##       Place Baker.pct Coakley.pct Baker.pct.margin
```

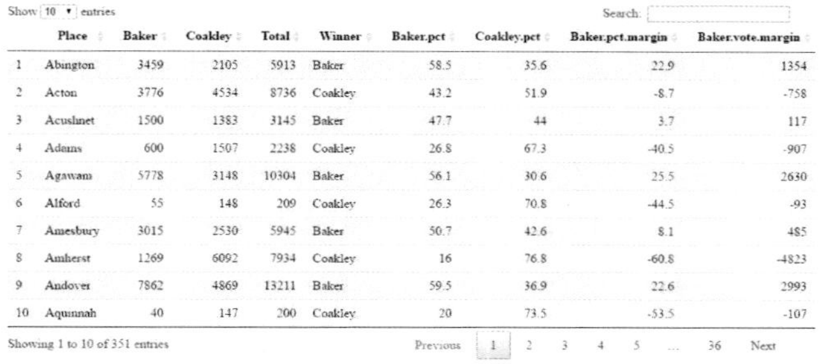

Show 10 entries Search:

	Place	Baker	Coakley	Total	Winner	Baker.pct	Coakley.pct	Baker.pct.margin	Baker.vote.margin
1	Abington	3459	2105	5913	Baker	58.5	35.6	22.9	1354
2	Acton	3776	4534	8736	Coakley	43.2	51.9	-8.7	-758
3	Acushnet	1500	1383	3145	Baker	47.7	44	3.7	117
4	Adams	600	1507	2238	Coakley	26.8	67.3	-40.5	-907
5	Agawam	5778	3148	10304	Baker	56.1	30.6	25.5	2630
6	Alford	55	148	209	Coakley	26.3	70.8	-44.5	-93
7	Amesbury	3015	2530	5945	Baker	50.7	42.6	8.1	485
8	Amherst	1269	6092	7934	Coakley	16	76.8	-60.8	-4823
9	Andover	7862	4869	13211	Baker	59.5	36.9	22.6	2993
10	Aquinnah	40	147	200	Coakley	20	73.5	-53.5	-107

Showing 1 to 10 of 351 entries Previous 1 2 3 4 5 ... 36 Next

Figure 12.5: A sortable, searchable table created with the DT package.

```
## 1 Westport       46.6        46.4              0.2
## 2   Natick       47.2        47.7             -0.5
## 3  Orleans       48.4        47.6              0.8
## 4   Milton       48.6        47.7              0.9
## 5 Carlisle       48.4        47.0              1.4
## 6  Wayland       48.8        47.4              1.4
```

Note that I got 6 rows back, not 5, because the towns of Carlisle and Wayland were tied at 1.4%. (I selected four columns here so there would be enough room to print out the important columns on this page.)

There are more of these types of highest and lowest results that would be interesting to see, but it starts getting tedious to write out each one. It feels a lot easier to do this by clicking and sorting a spreadsheet than writing out code for each little exploration. If you'd like to re-create that in R, you can view the winners data frame by clicking on it in the Environment tab at the top right, or running `View(winners)` in the console. Clicking on a column header once sorts by that column in ascending order; clicking a second time sorts the data frame by that column in descending order. This is a nice if unstructured way to view the data.

An even better way of doing this is with the DT package, which will create an interactive HTML table. Install it from CRAN, load it, and then run its datatable function on the data frame, just like this:

```
datatable(winners)
```

You'll get an HTML table that's sortable and searchable (see Figure 12.5). The table first appears in RStudio's viewer; but you can click the "Show in new window" icon (to the right of the broom icon) and the table loads in your default browser.

The DT package's Web site at https://rstudio.github.io/DT/ gives you a full range of options for these tables. A few I use very often:

- `datatable(mydf, filter = 'top')` adds search filters for each column
- `datatable(mydf) %>% formatCurrency(2:4, digits = 0, currency = "")` displays the numbers in columns 2:4 with commas (`digits = 0` means don't use numbers after a decimal point, and `currency = ""` means don't use a dollar sign or other currency symbol)
- `datatable(mydf, options = list(pageLength = 25))` sets the table default to showing 25 rows at a time instead of 10.
- `datatable(mydf, options = list(dom = 't'))` shows just the sortable table without filters, search box, or menu for additional pages of results – useful for a table with just a few rows where a search box and dropdown menu might look silly.

Although I've been using the DT package for years, I still find it difficult to remember the syntax for many of its options. Like with ggplot2, I solved this problem with code snippets, making it incredibly easy to customize my tables. For example, this is my snippet to create a table where a numerical column displays with commas:

```
snippet my_DT_add_commas
    DT::datatable(${1:mydf}) %>%
    formatCurrency(${colnum}, digits = 0, currency = "")
```

(If I have more than one numerical column, it's easy enough to replace one column number with several.) All my DT snippets start `my_DT_` so they're easy to find in an RStudio dropdown list if I start typing my_DT.

One more benefit of the DT package: It creates an R *HTML Widget.* This means you can save the table as a stand-alone HTML file. If you save a table in an R variable, you can then save that table with the htmlwidgets::saveWidget() function:

```
MA2014_results <- datatable(winners, filter = 'top') %>%
  formatCurrency(2:4, digits = 0, currency = "")

htmlwidgets::saveWidget(MA2014_results, file = "MA2014_results_table.html")
```

If you run that, you should see a MA2014_results_table.html file in your project's working directory. As with maps in the previous chapter, you can open this file in your browser just like any local HTML file. You can also upload it to a Web server to display directly or iframe on your website – useful for posting election results online.

That table also makes interactive data exploration easier. I can filter for just Baker's wins or Coakley's wins, sort with a click, use the numerical filters' sliders to choose small or large places, and more.

12.5 Visualizing election results

Is there a relationship between number of votes in a community and which candidate won? A scatterplot can help show trends. However, Boston is such an outlier population-wise, that it becomes difficult to see what's happening in the rest of the state (Figure 12.6).

```
ggplot(winners, aes(x = Total, y = Baker.pct.margin)) +
  geom_point()
```

One approach is to simply remove Boston to get a better look at trends (Figure 12.7):

```
maplot <- ggplot(winners[winners$Place != "Boston",], aes(x = Total, y = Baker.pct.margin))
+  geom_point()
maplot
```

Another approach is to use a logarithmic scale, which can be a more useful way of visually exploring relationships in data without having to toss out one or two extreme outliers. However, there are two challenges in this example. One is that you can only take logarithms of positive numbers, so I'd have to re-do the data structure instead of using a negative margin for places Baker lost. The other challenge is that if you're trying to tell a story to a mass audience, having to explain log scales in the midst of your election reporting may not be ideal. (That's not a problem when creating for-your-own-use data explorations, though.)

So I'll stick with the Boston-less scatter plot for now. If there's a pattern, it's not very visually dramatic. If you understand linear regression – finding a trend line of best fit among points – you can add a linear-regression line easily with ggplot's `geom_smooth(method="lm")`. (Caution: plotting a straight line isn't always the best choice for scatter plots; it depends on characteristics of your data.)

```
maplot +
  geom_smooth(method="lm")
```

Not much of a trend to see in Figure 12.8. Base R's cor() function can calculate the statistical correlations:

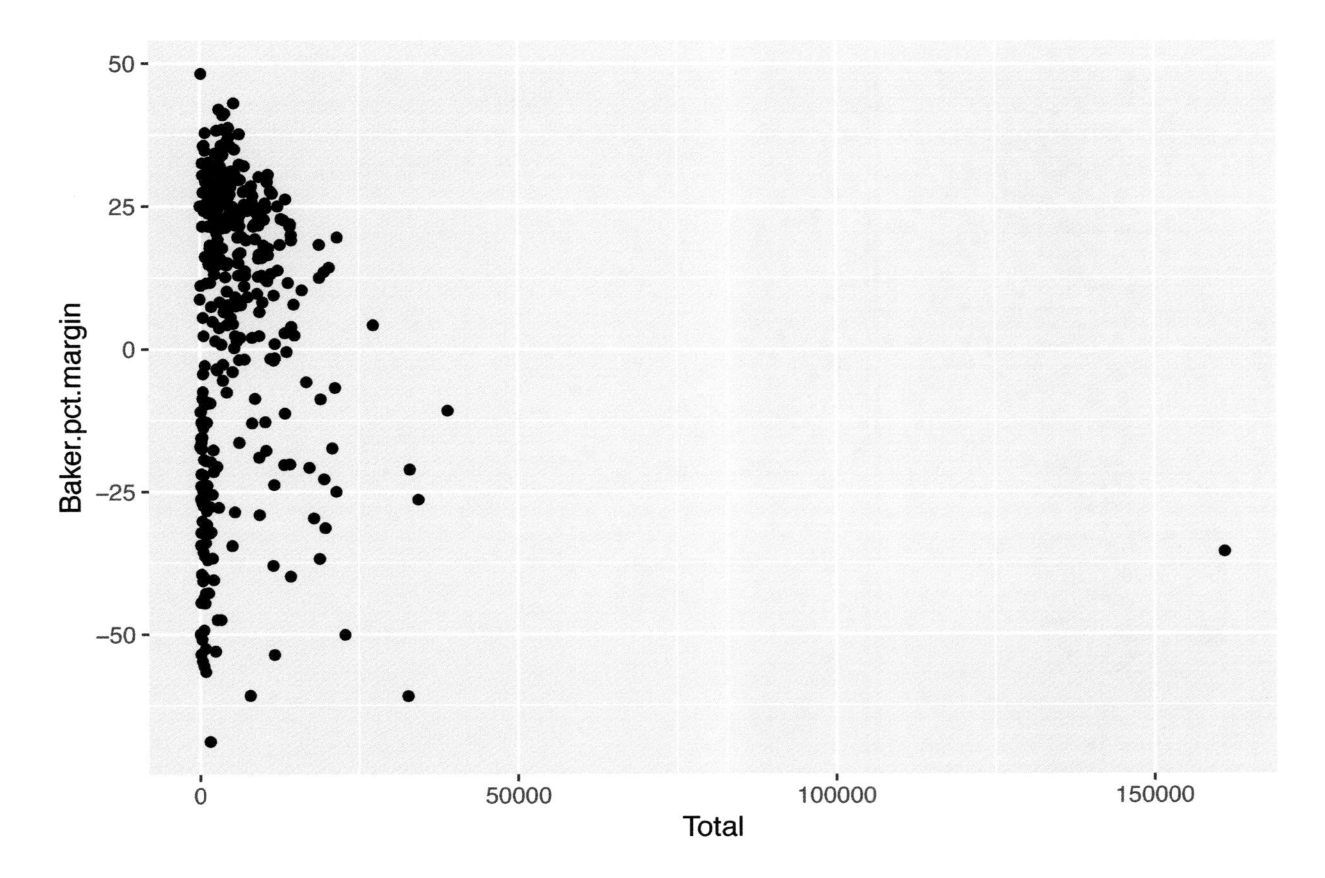

Figure 12.6: A scatterplot of population vs. percent victory margin.

```
cor(winners$Total, winners$Baker.pct.margin)

## [1] -0.08767372
```

`cor(winners$Total, winners$Baker.pct.margin)` will give more details, including statistics such as p-value and 95% confidence interval.

Finally, using geom_smooth() in ggplot2 without specifying lm (for "linear model") lets ggplot determine what type of model to use, as in Figure 12.9:

```
maplot +
  geom_smooth()
```

If you want to do more correlation calculations, there are several R packages that might be of interest. corrplot is designed to visualize correlations; the package includes an introductory vignette that can tell you more. The corrr package is billed as "a tool for exploring correlations," including rearranging correlations based on the strength of relationships.

12.6 Graph for a smaller set of results

This Massachusetts data had more than 350 results, meaning visualization choices such as bar graphs might not make sense. But when covering precinct results in one town, or results by town for just a handful of places, you might want a bar chart where the winner's bars are one color and the loser's are another.

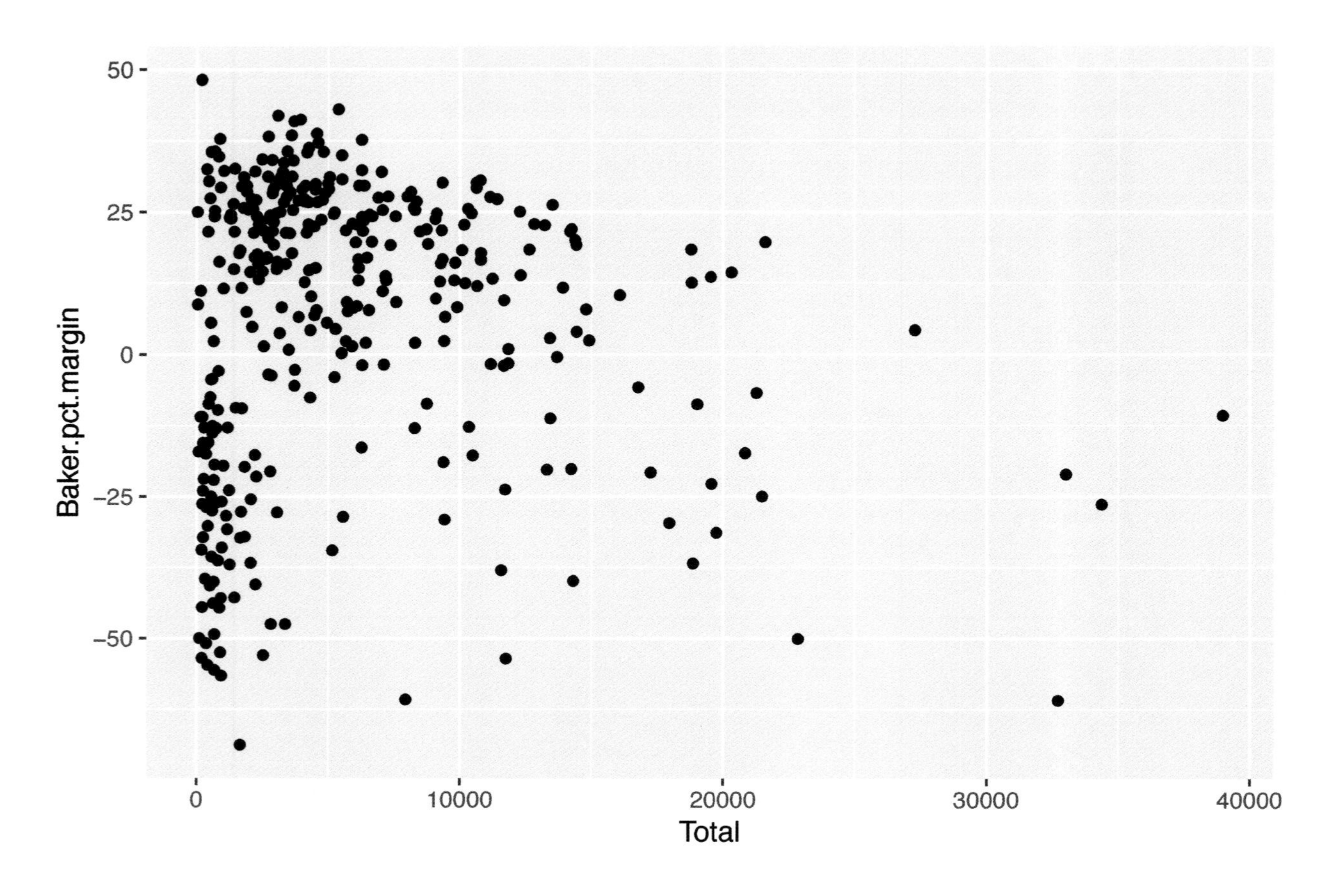

Figure 12.7: Scatterplot with outlier Boston removed.

This could also be useful for a local reporter covering a state-wide race who wants to examine one specific area's results, such as Cape Cod. The next code block subsets the data for towns on Cape Cod, and creates a bar graph with different colors for Baker and Coakley towns.

The first line defines all the towns I want in this Cape Cod graph and stores them in a variable `cape_cod_towns`. The second line keeps only those rows in the winner's data frame where the Place column is one of the cape_cod_towns, and stores them in a new variable called cape. Finally, the third statement creates a bar chart of Baker's winning percent margin in those towns, where the bar colors are based on the Winner column.

```
cape_cod_towns <- c("Falmouth", "Bourne", "Mashpee", "Sandwich", "Barnstable", "Hyannis",
                    "Yarmouth", "Dennis", "Harwich", "Brewster", "Chatham", "Orleans",
                     "Eastham", "Wellfleet", "Truro", "Provincetown")

cape <- filter(winners, Place %in% cape_cod_towns)

ggplot(cape, aes(x=Place, y=Baker.pct.margin, fill=Winner)) +
  geom_col() +
  labs(title = ma_gov_headline, caption = ma_gov_datasource) +
  theme(axis.text.x = element_text(angle = 45, vjust = 1.2, hjust = 1.1))
```

This ggplot2 (Figure 12.10) happens to set the Republican candidate in red and Democrat in blue, but that won't always be the case. You can set your own fill color scheme with scale_fill_manual() using syntax something like `scale_fill_manual(values=c("red4", "blue4"))`.

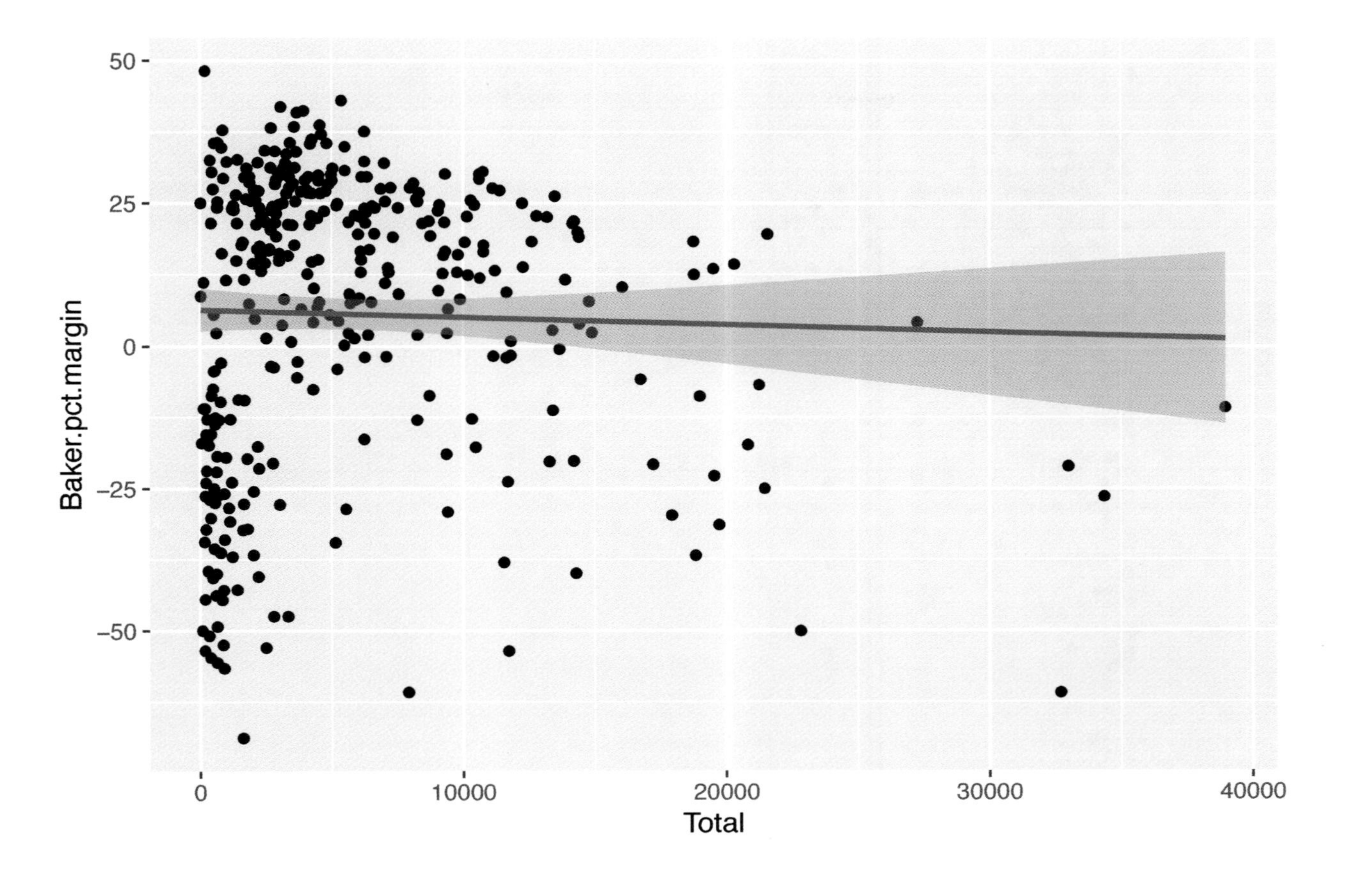

Figure 12.8: Scatter plot with linear regression line.

There are other exploratory graphics you might want to generate from election results. If you're covering a city election with 15 or 20 precincts, a bar chart of margins per precinct might be interesting. In a tight election, you might want to examine totals for third-party candidates and blanks versus the winning candidate's margins. This is just a starting point for you to create your own election recipes, based on topics that might be interesting in the area that you cover.

12.7 plotly

As discussed in Chapter 9, one of the easiest ways to make interactive graphs is with the plotly package. If you already know ggplot, plotly's ggplotly() function can turn a static ggplot into a JavaScript interactive graph.

Let's go back to the initial scatterplot, looking at vote total versus Baker's percent margin of victory. I'll tweak it slightly by adding a *label* option as part of the aes(), and then save the plot to a variable.

```
library(ggplot2)
scatplot1 <- ggplot(winners, aes(x = Total, y = Baker.pct.margin, label = Place)) +
  geom_point()
```

Why add the Place column as a label option? For an interactive version of this scatterplot, I'd like to be able to mouse over a point and see the city/town, not only the vote total and margin of victory. If I don't add the Place column in the plot, it can't show up in the tooltip.

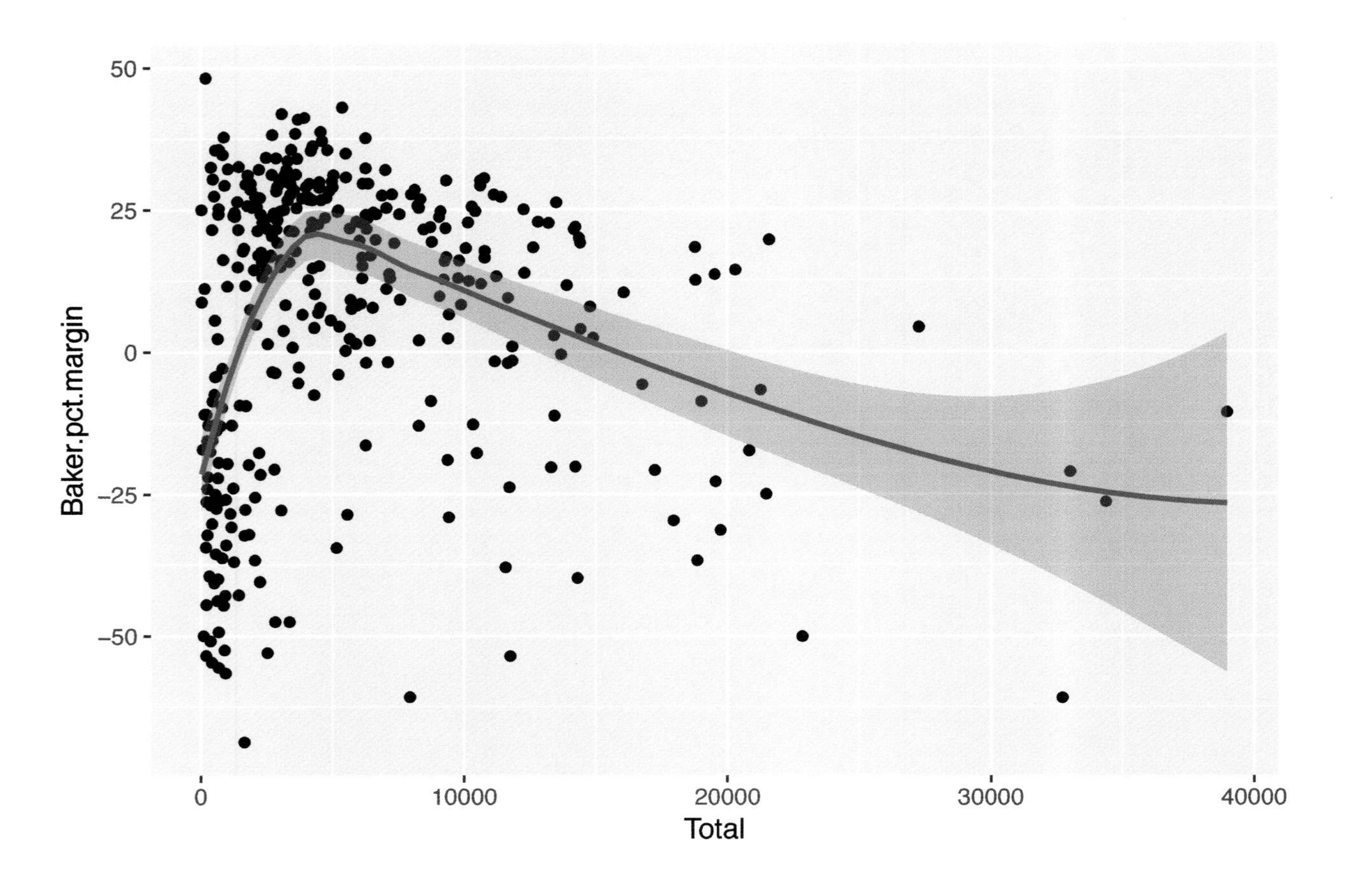

Figure 12.9: Letting R pick the model.

Now if I run ggplotly on that scatterplot, I'll get an interactive version allowing me to mouse over any point to see underlying data (Figure 12.11).

```
library(plotly)
ggplotly(scatplot1)
```

This interactive version also allows users to click and drag a portion of the plot in order to zoom in – one way to examine the data excluding Boston, or zero in on areas with a lot of points clumped together.

12.8 Other interactive alternatives

12.8.1 taucharts

taucharts is my favorite package for interactive scatter plots for two reasons: tooltips automatically include every variable in your data frame, not just the two being plotted; and you can add a dropdown menu to choose different models for trend lines.

This code:

```
tauchart(winners) %>%
tau_point("Total", "Baker.pct", color="Winner") %>%
tau_guide_x(label="Total Votes") %>%
tau_guide_y(label="Baker Pct") %>%
```

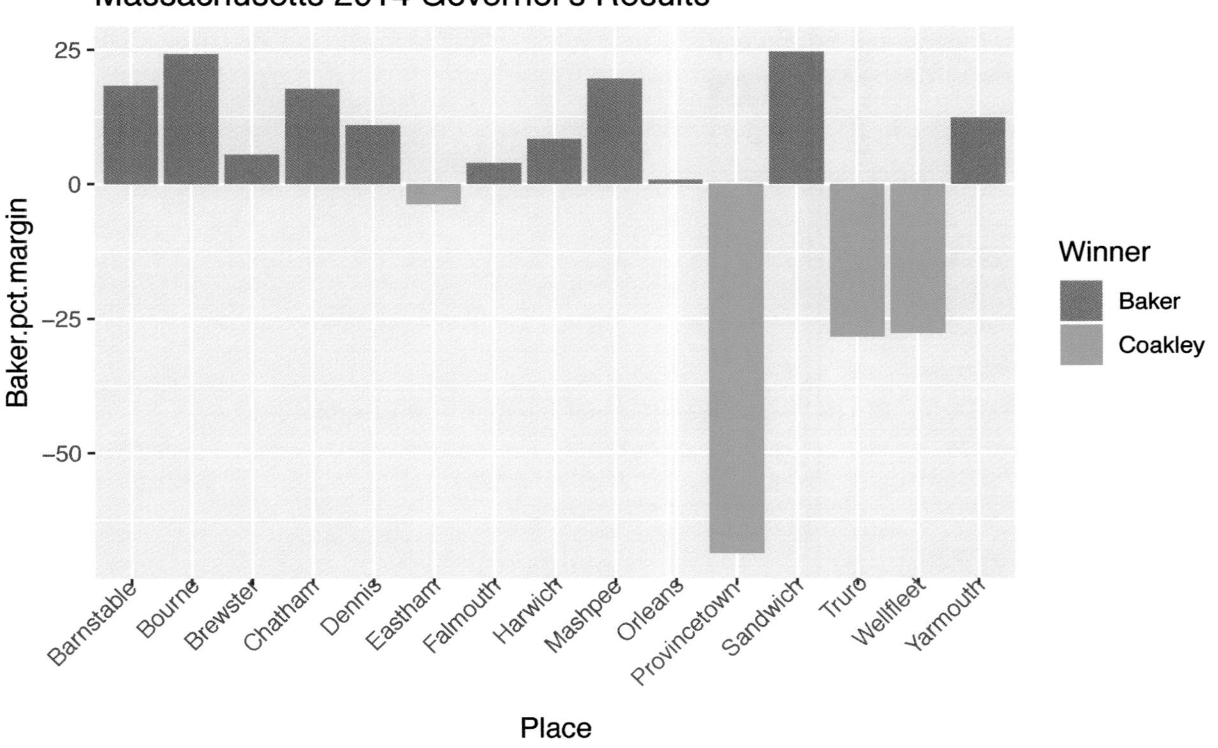

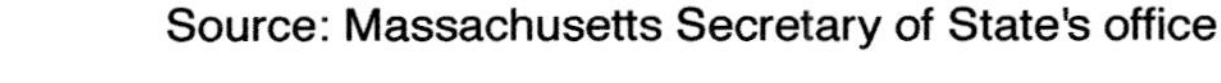

Figure 12.10: Bar chart with winners and losers by town.

```
tau_tooltip() %>%  # includes all variables in mydf
tau_trendline( showPanel = TRUE )  %>%
tau_title("2014 MA Governors Results")
```

Produces a graph like Figure 12.12.

I've had occasional problems viewing taucharts graphs in the RStudio viewing panel, but clicking the "show in new window" icon above the panel to open the visualization in a browser usually works fine. See more about the taucharts R package at http://rpubs.com/hrbrmstr/taucharts.

12.8.2 highcharter

Highcharter is an R wrapper to one of my favorite JavaScript libraries, Highcharts. Both the original library and R package are well documented, and the graphics make for publication-quality visualizations. Note, however, that Highcharts.js is only free for personal and non-profit projects; government and commercial use require a paid license (see more at highcharts.com). The highcharter package can be installed and loaded like any other from CRAN with install.packages() and library() or pacman::p_load().

highcharter's hchart() function is similar to ggplot2 graphing in that "You pass the data, choose the type of chart and then define the aesthetics for each variable," package creator Joshua Kunst explains on the package's website. See more at jkunst.com/highcharter.

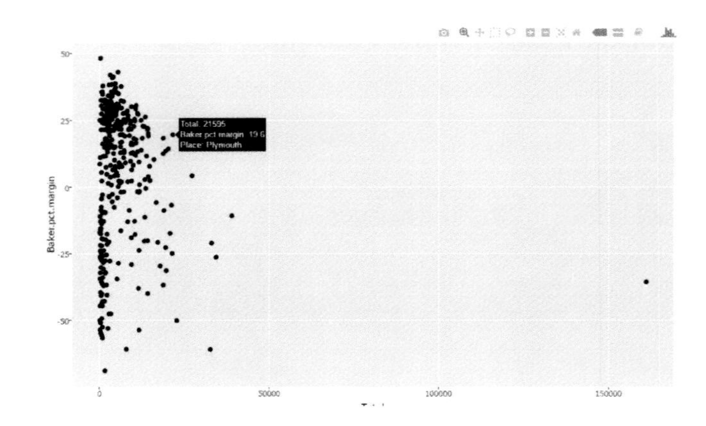

Figure 12.11: Making a scatter plot interactive with ggplotly().

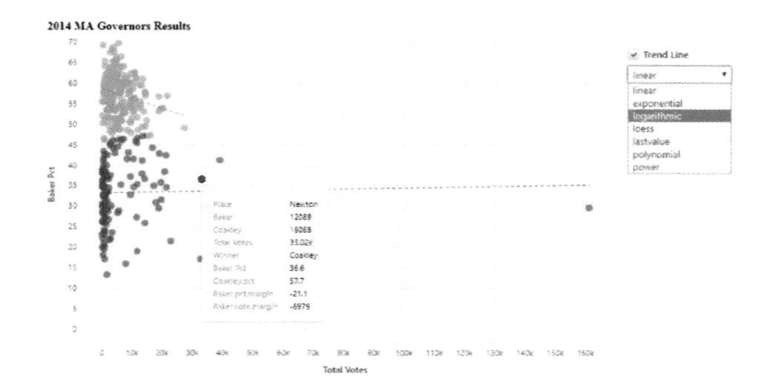

Figure 12.12: An interactive scatter plot created with taucharts.

12.8.3 metricsgraphics

This R wrapper and htmlwidgets implementation of the MetricsGraphics.js library can create interactive line charts, bar charts, and scatterplots. More information is available at the package's website, http://hrbrmstr.github.io/metricsgraphics/.

To see other interactive R package options, check out the HTML widgets gallery at http://gallery.htmlwidgets.org.

12.9 Wrap-up

We covered processing raw election data to find winners, including functions in my rmiscutils package. We also got a look at a new pipe operator in the magrittr package, saving and loading data in Rda format, creating interactive tables with the DT package, and plotting and calculating correlations. Finally, we took a look at several interactive data visualization packages in addition to plotly.

Next up: Dealing with dates.

12.10 (Non-election) inspiration

If you'd like to see some of these skills in action in a non-electoral context, install the fivethirtyeight package with `install.packages("fivethirtyeight")`, load it, and then view the *bechdel vignette* with `vignette(topic = "bechdel", package = "fivethirtyeight")`. This will show you analysis for the FiveThirtyEight.com story "The Dollar-And-Cents Case Against Hollywood's Exclusion of Women," in which Walt Hickey shows that movies which feature three-dimensional female characters have a greater return on investment in the U.S. than other types of films.

12.11 Additional resources

If you're working with large data files, you may want to look into alternatives to base R's save() and load() functions. Several packages aim to make it faster to store and load R objects, including `fst` and `feather` (feather is also useful for those who know Python as well as R, since that binary file format can be read by both languages). Check out the packages on CRAN for more information.

For analyzing and visualizing pre-election polling data in R, the pollstR package is an R client for the Huffington Post's Pollster API. This source has mostly data on U.S. contests, although it occasionally includes data from other major elections worldwide, such as the 2017 France presidential race. https://github.com/rOpenGov/pollstR

Heat maps can be an interesting way to visualize changes in results over time. Peter Aldhous, a science reporter with BuzzFeed News and investigative reporting instructor at the University of California Santa Cruz, posted materials from his National Institute for Computer-Assisted Reporting training session that includes creating a heat map with ggplot2. http://paldhous.github.io/NICAR/2017/r-analysis.html

Interested in visualizing election results by party for a legislature such as the U.S. Senate or U.K. House of Commons? Check out the ggparliament package on GitHub at https://github.com/robwhickman/ggparliament

My guide to Election Night resources for the 2016 election includes a link to compare forecasts with results, and how to use the pollstR package to pull data from the Huffington Post's Pollster API. http://www.computerworld.com/article/3139884/data-analytics/r-resources-for-election-night.html

Kan Nishida has a more stats-heavy example of using R to analyze election results, using techniques such as K-means clustering to see which California counties are most similar to each other based on 2016 election results. https://bit.ly/Rsimilarities

Chapter 13

Date calculations

When is a crucial part of journalism's classic Who, What, Where, When, and Why? But in data analysis, you often want to do more with dates than just report when something happened (or is expected to happen). Date arithmetic – calculating the time between events – can also be an important part of a story.

For example, NBC News used date arithmetic for its investigation of bridge inspections after the 2007 collapse of a bridge in Minneapolis. Federal regulations require bridges to be inspected every two years, but data showed that many bridges went longer between check-ups (NBC's analysis wasn't necessarily done with R, but it could have been. You can see their series at http:/bit.ly/NBCbridges.)

13.1 Project: New York City restaurant inspections

In this chapter, we'll start off with some basics of using dates in R. Then, we'll take a look at New York City public restaurant-inspection data, calculating **how long it takes for follow-up inspections after a restaurant is cited for a critical violation.**

And, we'll work through a real-world dilemma where data needs to be reformatted.

If you're interested in trying out similar date skills on U.S. bridge or dam data instead, files already formatted for easier analysis can be purchased from the National Institute for Computer-Assisted Reporting's Database Library at http://ire.org/nicar/database-library/. NICAR is part of Investigative Reporters and Editors.

13.2 What we'll cover

- Turning a string like "6/27/2019" into an R date object
- Doing date calculations with both base R and the lubridate package
- Finding prior and next values with dplyr's lead() and lag()
- Dealing with times

13.3 Packages needed in this chapter

```
pacman::p_load(lubridate, janitor, ggplot2, dplyr, rio)
```

You'll also need the Hmisc package, but I suggest installing it if you don't already have it, but *not* loading it. Its summarize() function can conflict with dplyr's summarize(), and I'd like to avoid having to write out `dplyr::summarize()` numerous times. In fact, if you've previously loaded Hmisc, it's worth specifically *unloading* it from memory with the rather unintuitive unloadNamespace() function:

```
unloadNamespace("Hmisc")
```

13.4 Get started with dates in R

In R, as in most programming languages, there's a difference between a character string that *looks like* a date – "2019-06-21" or "June 21, 2019" – and an actual date *object* with specific methods (class-specific functions) that only work on dates.

A date object can print out as "2019-06-21", but its behavior will be different from the string version that also prints out as "2019-06-21". For example, `"2019-06-21" + 1` throws an error if "2019-06-21" is a character string, but will return "2019-06-22" for a date.

13.4.1 How to create date objects

You can create a date object from a string by using R's `as.Date()` function. `as.Date()` expects dates in yyyy-mm-dd or yyyy/mm/dd format.

`as.Date("2019-06-21")` will create an R date object for June 21, 2019.

But what happens when you've got a date in typical American mm/dd/yyyy or European dd/mm/yyyy format?

There are a couple of ways to turn those into date objects. The easiest way is with the lubridate package. Its mdy() function will convert strings in a lot of different month/day/year formats, including `mdy("6/21/2019")`, `mdy("6-21-19")`, and `mdy("06212019")`. There are similar functions for day/month/year `dmy()` and yyyy-mm-dd `ymd()`.

Base R's method is a bit more complicated. I recommend lubridate unless you are coming from another programming environment and are familiar with something called the POSIX (Portable Operating System Interface for Unix) standard. The help file for R's strptime() function, `?strptime` , includes help with POSIX formatting.

13.4.2 Simple (date) arithmetic

Regular `+` and `-` operators work on date objects as well as numbers. `mdy("6/21/2019") + 1` returns a date object one day later than 6/21/2019.

As mentioned in the chapter on writing your own functions, `Sys.Date()` returns the current date. So, `today <- Sys.Date()` stores the current date in a variable called today.

Once you've defined today, `today + 1` gives you tomorrow's date, while `today - 1` gives you yesterday's date.

lubridate makes it easy to add or subtract years, months, or weeks, too. You can't just add 365 days to a date object to get the next year, thanks to the one-in-four chance you need a leap year's 366 days. Months are even more erratic, having anywhere from 28 to 31 days. A date operation has to understand what a "year" or "month" is for each specific period.

The easiest way to do that kind of date arithmetic is with lubridate's years() and months() functions. `today + years(2)` adds two years to `today`, while `today - months(1)` returns a date that's a month earlier than `today`. There's also weeks(), such as `today + weeks(5)` to add 5 weeks to today's date.

The difference between two dates. For the NYC restaurant inspection project, we'll want to know the number of days between two dates. If you subtract one date object from another one, you get a `difftime` object in return. For example, how many days between "today" (the system date) and June 21, 2019?

```
mdy("06/21/2019") - Sys.Date()
```

```
## Time difference of 366 days
```

Don't forget to load the lubridate package first if you want to use mdy() and associated functions, either with library(lubridate) or pacman::p_load(lubridate).

If you just want an *integer* with the number of days, and not a difftime object, as.numeric() will turn it into a number:

```
as.numeric(mdy("06/21/2019") - Sys.Date())
```

```
## [1] 366
```

If the second item is later than the first, the difference is a negative number.

Base R's difftime() *function* finds the difference between two dates, using the format `difftime(dateobject1, dateobject2)`. The advantage of using `difftime()` instead of a simple `-` is that you can specify the units, such as "days", "weeks", or "secs", such as `difftime(mdy("06/21/2022"), Sys.Date(), units = "weeks")`.

13.5 Get NYC inspection data

When I created this project, New York City had posted its restaurant inspection data at NYC Open Data, https://opendata.cityofnewyork.us/. You can search that site for restaurant inspections to get the latest data. Or, to make sure we're using the same data set, you can use the files from the book GitHub repo.

I downloaded the three inspection files – the zipped csv data file, data dictionary (explaining each column) spreadsheet, and additional explanations (Word doc) – to a nycinspections subdirectory. After reading explanations about the data and its columns, I'll import the data file with `inspections <- rio::import("nycinspections/nycdata.zip")` and examine the structure with dplyr's glimpse() function. rio handles the unzipping before importing data.

When I ran `rio::import()`, all the columns were imported as characters except the first ID column named CAMIS, which came in as integers. However, there are a couple of issues with these column types. If I'm going to group the data by each restaurant, I want IDs to be categories, which means I need character strings or factors. And, if I want to do date calculations, I'll need the dates as R date objects.

Turn the CAMIS ID column into strings with `inspections$CAMIS <- as.character(inspections$CAMIS)`. Or, if you want to follow best practices, create a new id column from the CAMIS column with `inspections$id <- as.character(inspections$CAMIS)`.

Some of these columns' names have spaces, which can cause complications in R. I'd suggest running `inspections <- janitor::clean_names(inspections)` to generate R-friendlier column names.

Next, create a new column of date objects from the `inspections$inspection_date` data frame column using lubridate's `mdy()`:

```
inspections$date_inspected <- mdy(inspections$inspection_date)
```

13.5.1 Data prep

Before writing any more code, this would be a good time to pause to think about how the data needs to be structured for the question we have. We want to analyze time between restaurant inspections after a critical violation has been discovered.

This series of data-wrangling steps might *seem* like enough: Group data by restaurant ID, sort by date from oldest to newest, and calculate the number of days between each row's inspection date and the one before it. Then, filter for rows where critical violations were found, in order to have number of days between a critical violation and the next inspection.

However, that plan has a problem: *This data set can have multiple rows for a single inspection.* In other words, each row doesn't represent one *inspection*; it represents a single *violation*, and one inspection can have multiple violations. So, a row after a critical violation could be an inspection on the same date, as in this portion of the inspections data frame:

dba	critical_flag	date_inspected
MORRIS PARK BAKE SHOP	Critical	2016-02-18
MORRIS PARK BAKE SHOP	Not Critical	2016-02-18

Those two lines would show zero days between an inspection with a critical violation and the next inspection. But of course, that next *row* isn't the next *inspection.*

One way to deal with this complication is to group by restaurant ID *and date*, and then add a column to see whether *any* violation in the group was critical.

I've already covered how to group and sort with dplyr, using the group_by() and arrange() functions:

```
inspections %>%
  group_by(id, date_inspected) %>%
  arrange(id, date_inspected)
```

This data set includes a critical_flag column. Running `table(inspections$critical_flag)` shows there are three possible values in that column: "Critical", "Not Critical", and "Not Applicable".

To find out whether *any* of the violations in a group are "Critical", base R's any() function is perfect. It answers the question: Are any of the logical tests on a vector TRUE?

Here's how it works. Let's create a simple vector: `myvector <- c("Not Critical", "Not Critical", "Critical", "Not Applicable")`. Are any of those values "Critical"? any() uses this syntax: `any(myvector == "Critical")` which asks "Are any of the values in myvector equal to 'Critical'?"

Try similar code on a vector which doesn't contain "Critical", such as:

```
myvector2 <- c("Not Critical", "Not Critical", "Not Applicable")
any(myvector2 == "Critical")
```

```
## [1] FALSE
```

any() only returns one value. To *vectorize* (get a vector in return) and run this operation over many values, `ifelse()` will be useful. Review: `ifelse()` uses the format `ifelse(logical test, resulting value if the test is TRUE, resulting value if the test is FALSE)` . For this problem, I want to create a new column called found_critical which is TRUE if any violations in the group are critical, and FALSE otherwise.

Here's my full code, with ifelse() as the final step, grouping by a few extra columns:

```
inspection_critical_test <- inspections %>%
  group_by(id, date_inspected, dba, boro) %>%
  arrange(id, date_inspected) %>%
  summarize(
     found_critical = ifelse(any(critical_flag == "Critical"), TRUE, FALSE)
  )
```

Why add dba and boro in the groups? I didn't necessarily *need* to group by the dba (restaurant name) and boro (county) columns in order to do my analysis. However, I added those columns in group_by() because I'd like to include those columns in my new data frame. Without explicitly mentioning them, they won't appear in the new, summarized data frame.

The next step is finding the number of days between inspections. Subtracting two date objects will find the difference. But how best to locate the "previous date" for each date?

13.5.2 dplyr's lead() and lag()

lead() and lag() make it easy to calculate "next" and "previous" values in a vector. To find the prior value for each item in a vector, use the format `lag(myvector)`. If you want to find the values *two* before each item in a vector, run `lag(myvector, n = 2)`. lead() is similar, but looks at the next value(s) instead of previous ones.

You can add a new column to calculate the difference between each value and the one before using lag() and the format

`new_column <- as.numeric(datecolumn - lag(datecolumn))`.

Note that lag() and lead() also work for calculating differences between a vector of *numbers* and items before or after each one, not just dates.

One final point before using lag() on the inspection data: The data is currently grouped by id *and date inspected.* Any operation run on this grouped data will be run within each group, and each group only has one inspection date. That's not helpful. Instead, we'd like to calculate time between inspections *for each restaurant id only.*

We can fix this by ungrouping the data with the ungroup() function, and then re-grouping by restaurant id alone.

Here's the code:

```
inspection_critical_test <- inspections %>%
  group_by(id, date_inspected, dba, boro) %>%
  arrange(id, date_inspected) %>%
  summarize(
     found_critical = ifelse(any(critical_flag == "Critical"), TRUE, FALSE)
  ) %>%
  ungroup() %>%
  group_by(id) %>%
  mutate(
    days_since_last_inspection = as.numeric(date_inspected - lag(date_inspected))
  ) %>%
  filter(found_critical == TRUE)
```

Let's go over this more-complex-than-expected real-world example:

- The first line creates a new data frame from inspections data.
- Line 2 groups the data by restaurant id and inspection date, which we need for our analysis, as well as restaurant name (dba) and borough so those will appear in the summarized data frame. Additional functions will be applied within each group.
- Line 3 sorts data by restaurant id and inspection date.
- summarize() creates a column that's TRUE if the inspection for a restaurant on that date found a Critical violation and FALSE if not.
- Next lines ungroup the data, re-group it by id only, add a column that calculates the number of days from that inspection date to the previous one, and then filter for only inspections where a Critical violation was found.

Now you can analyze times between inspections using tools discussed in other chapters, such as basic summaries with Hmisc::describe() and base R's hist() for a histogram.

```
Hmisc::describe(inspection_critical_test$days_since_last_inspection)

## inspection_critical_test$days_since_last_inspection
##        n  missing distinct     Info     Mean      Gmd      .05      .10
##   102580    21090      618        1    156.9    151.3       14       17
##      .25      .50      .75      .90      .95
##       28      138      239      374      399
##
## lowest :    1    2    3    4    5, highest:  873  915  981  984 1071

hist(inspection_critical_test$days_since_last_inspection)
```

If I were a reporter working with this data, I'd want to know more about why some restaurants were showing more than 2 years between finding of a critical violation and a re-inspection. Is it simply that the data is incomplete? Are some "critical" problems not really all that problematic? Or is there something potentially newsworthy in this inspection data? It would be worthwhile to do more reporting on this data set.

13.5.3 More date functions worth knowing

Before wrapping up our date chapter, I'd like to outline a few more useful things to know about dates and times in R, some of which have been touched on earlier:

You can *find the day of the week* for any date object with base R's weekdays() function. `weekdays(my_date_object)` gives the full weekday name, such as "Monday" or "Tuesday". `weekdays(my_date_object, abbreviate = TRUE)` returns an abbreviated version.

It can be helpful to **categorize dates by week, month, quarter, or year.** Base R's cut() function is designed to put numbers into categories, but it has some special and somewhat hidden powers when used with dates. As mentioned in Chapter 10, you can cut date objects by week, month, quarter, or year. `cut(my_date_object, breaks = "month")` or just `cut(my_date_object, "month")` will return the first day of the month for that date, but as a *factor*, not a date object. `as.Date(cut(my_date_object, "month"))` will return a date object. Breaks of "week", "quarter", and "year" work similarly, with week offering the choice of start.on.monday = TRUE or start.on.monday = FALSE (in which case the week starts on Sunday).

lubridate can generate date categories with its floor_date() function. floor_date() takes two arguments: a date object and the desired unit: week, month, bimonth, quarter, halfyear, or year. And, it returns a date object, not a factor. So, `floor_date(my_date_object, "month")` will return the first of the month for that date.

In addition, lubridate has separate functions week(), month(), quarter(), and year(). You'll get back integers from these functions – for example, `week(as.Date("2019-06-21"))` will return the number 25, for the 25th week of the year, not "2019-06-17" for the first day of that date's week.

There might be times when you want to generate a sequence of dates, something like "the first of every month starting with January 1, 2019". As mentioned in Chapter 10, `seq.Date()` will do this for you. It takes three arguments: a date object, how many elements you want in your vector of date, and your desired interval: `seq.Date(my_date_object, length = vectorlength, by = "units")`.

Here's how to get the first day of each quarter in 2019:

```
seq.Date(mdy("1/1/19"), length = 4, by = "3 months")

## [1] "2019-01-01" "2019-04-01" "2019-07-01" "2019-10-01"
```

13.5.4 Dates with time of day

Sometimes you don't just need the date, but you also need – or have – the time of day specifying hours, minutes, and perhaps seconds. R has a couple of date-time classes (think of classes as types of objects if you're not familiar with object-oriented programming): `POSIXlt` and `POSIXct`.

I don't have space to go into detail on these, but I do want to warn you that the difference between the two can confuse beginners and trip up even more experienced R users.

In brief: An object of the *POSIXct* class stores the number of seconds since Jan. 1, 1970. Dates after then are a positive number; dates before then, a negative number. (This isn't a quirk of R, but a legacy from the early days of computing and the Unix operating system.) It prints out looking like a date object, but with time in hh:mm:ss format and a time zone added.

A *POSIXlt* object is a list of vectors with the date/time's seconds, minutes, hour, day of the month, month, year, day of week, day of year, and whether Daylight Savings Time is on (along with optional time zone and GMT offset).

It's possible for an R object to have characteristics of both POSIXct and POSIXlt classes. Run `class(Sys.time())` and you'll see.

You can perform date arithmetic on date-time objects, and use functions like cut.Date() – with sec, min, hour, and day as breaks as well as week, month, quarter, and year – and floor_date().

One of the most important things to know about these is that some functions and data structures only work with one date-time class while others can use either. For example, the R Date-Time Classes documentation suggests that "'POSIXct' is more convenient for including in data frames, and 'POSIXlt' is closer to human-readable forms." (You can read the documentation by running `?DateTimeClasses` in your R console.)

If you're having trouble dealing with date-time objects in R, the culprit may be using POSIXlt when you need POSIXct or vice versa. Reading a function's documentation to see what type of object it needs as input or what it's generating as output can sometimes help.

13.6 Wrap-up

We covered creating date objects, adding and subtracting dates, finding the difference between two dates, using dplyr's lag() to get the difference between one item and the prior item in a vector, and adding times to date objects.

Next up: Massaging and manipulating text

13.7 Inspiration

The Stanford Open Policing Project collects data from states throughout the U.S. on police traffic stops. The project posted a tutorial on analyzing Connecticut data with R, including use of lubridate and dplyr, at http://bit.ly/TrafficStopTutorial.

NY Times restaurant inspection interactive map:

http://www.nytimes.com/interactive/dining/new-york-health-department-restaurant-ratings-map.html

13.8 Additional resources

Video lecture on dates in R by Dr. Roger Peng for a Coursera class https://bit.ly/RDatesTimes.

If you need to work with times alone, without dates attached, the tidyverse includes an hms package. See more at the package's GitHub repo: https://github.com/tidyverse/hms.

Chapter 14

Help! My data's in the wrong format!

Anyone who's worked with data knows that sometimes, data isn't just messy; it's in a format that's downright analysis-hostile. But with R packages like tidyr (or the earlier reshape2) and dplyr, headache-inducing spreadsheets can be wrangled into shape.

14.1 Project: Election results in a PDF

In this chapter, we'll look at election results in a less-than-ideal format and turn them into 'tidy' format for easier analysis. In fact, we'll start off with results in a PDF!

14.2 What we'll cover

- Converting a PDF to Excel
- Reshaping data into analysis-friendly tidy format
- Finding "top 2" results in a group
- Adding rankings from low to high or high to low.

14.3 Packages needed in this chapter

```
pacman::p_load(tidyr, dplyr, janitor, readxl)
```

You may also want the pdftables and/or tabulizer packages, more on them soon.

14.4 Human vs. machine optimizing

Back in Chapter 6, we discussed a 'tidy' data format that has one observation, or record, in each row. However, what's optimal for a machine isn't always the most *human-friendly* of data formats.

The table in Figure 14.1 is probably one of the most common and easy-to-digest format for viewing election results:

That's easy to scan, but not necessarily "tidy". One issue: Some important information is *in column names* instead of within the data, such as candidate names. As we saw in Chapter 12, if the question is "which

	A	D	E	F	G	H	I	J	K
1	City/Town	Baker/ Polito	Coakley/ Kerrigan	Falchuk/ Jenni	Lively/ Sa	Mccormic	All Others	Blank Vot	Total Votes
2	Abington	3,459	2,105	203	50	45	4	47	5,913
3	Acton	3,776	4,534	242	39	59	15	71	8,736
4	Acushnet	1,500	1,383	71	90	46	2	53	3,145
5	Adams	600	1,507	48	27	33	0	23	2,238
6	Agawam	5,778	3,148	837	206	123	10	202	10,304
7	Alford	55	148	1	1	1	0	3	209
8	Amesbury	3,015	2,530	226	42	52	9	71	5,945
9	Amherst	1,269	6,092	302	63	42	19	147	7,934
10	Andover	7,862	4,869	270	61	57	2	90	13,211
11	Aquinnah	40	147	6	3	3	0	1	200
12	Arlington	6,243	12,435	622	104	103	13	221	19,741
13	Ashburnham	1,205	831	87	31	21	1	18	2,194
14	Ashby	754	443	68	19	10	0	10	1,304
15	Ashfield	200	587	47	2	6	1	24	867
16	Ashland	2,954	2,523	152	30	21	3	53	5,736
17	Athol	1,884	1,266	161	70	29	0	63	3,373
18	Attleboro	6,652	4,953	289	153	123	4	121	12,295
19	Auburn	3,743	2,250	174	56	55	3	63	6,344

Figure 14.1: Election results table in non-tidy format.

City Election Official Results - November 7, 2017

	1	2	3	4	5	6	7	8	9	10	11	12	13	14	15	16	17	18	Total
Mayor (1 seat) for 4 years																			
Blanks	7	2	2	3	2	9	2	7	2	2	0	4	1	2	1	1	1	0	48
John A. Stefanini	603	630	384	585	517	540	402	396	254	132	673	226	286	138	310	96	68	237	6457
Yvonne M. Spicer	755	844	657	873	708	748	694	682	371	250	747	335	438	231	274	153	134	234	9128
Write-ins	8	2	4	8	5	8	4	7	10	3	11	4	6	6	6	0	4	5	101
At Large City Councilor (2 seats) for 2 years																			
Blanks	642	660	464	716	513	604	543	591	310	181	658	279	372	208	349	172	126	293	7679
George P. King, Jr.	638	599	485	611	599	620	489	437	299	162	635	234	302	144	230	93	83	195	6865
Pablo Maia	89	104	107	103	95	93	90	108	93	37	127	75	101	65	118	51	46	83	1583
Christine A. Long	645	808	487	583	594	599	379	459	248	167	602	228	292	148	249	82	93	201	6844
Cheryl Tully Stoll	724	775	520	917	660	686	687	575	320	224	835	318	391	185	230	102	66	178	8393
Write-ins	8	10	11	8	3	8	6	16	4	3	7	4	4	4	6	0	0	2	104
District 1 City Councilor (1 seat) for 2 years																			
Blanks	81	77																	158
Charlie Sisitsky	710	682																	1392
Joseph C. Norton	579	713																	1292
Write-ins	3	6																	9
District 2 City Councilor (1 seat) for 2 years																			
Blanks			125		124														249
Pam Richardson			574		667														1241
Jeanne I. Bullock			321		431														752
Write-ins			7		10														17
District 3 City Councilor (1 seat) for 2 years																			
Blanks				163			99												262
Adam C. Steiner				888			616												1504
Joel Winett				412			386												798
Write-ins				6			1												7

Figure 14.2: A look at a PDF of election results.

candidate got the most votes?", the *number of votes* is within the data; but *names of candidates* are each in a column name.

There are ways around this, but the code can get complicated. Tidy data can simplify your scripts. In this chapter, we'll look at a City Council race where the top *2* vote-getters win seats.

Let's get started.

14.5 The raw data

Figure 14.2 gives a look at general election results for Framingham, Massachusetts in 2017. The Town Clerk's office released election results as a PDF. Had they been in a spreadsheet, the mayoral results wouldn't be too tough to handle. The At-Large City Council race, which we'll be examining here, is more challenging. The PDF shows the two overall winners in bold but not the top candidates by precinct.

First, though, we need to get the data out of that PDF.

14.6 Extracting data from PDFs

There's an R package on CRAN, pdftables, that can extract tables from PDFs. However, you have to sign up for an API key at pdftables.com, and after converting 50 PDF pages for free, you need to pay for the service. It's not very expensive for occasional use – a credit for 500 pages that's good for a year only costs $15 – but that may make it less useful if your goal is reproducible research for your audience.

The rOpenSci project's tabulizer package is an excellent choice for extracting reasonably well structured tables from PDFs. It's definitely worth adding to your R toolkit, with the format `mydata <- extract_tables(myfile, output = "data.frame")`. However, for this particular nightmare of a data dump, you'll end up with the data spread out over two data frames with different structures, and some tweaking is required. I'll show you the code for that at the end of this chapter.

For now, though, I'd like to get started on working with the actual data instead of getting sidetracked on the PDF problem. For occasional use on especially bad PDF data, I like the CometDocs service at cometdocs.com. It offers five conversions free each week for registered users and also allows people to upload and convert a document without registering. And, journalists who are members of Investigative Reporters and Editors have been eligible for additional conversions without paying for a subscription (check with IRE to see if that offer still holds). While this sacrifices R-from-start-to-finish reproducibility, I'm willing to begin with an Excel equivalent of the PDF and write my R code from there.

If you want to use pdftables, though, the code to convert the PDF to Excel is:

`convert_pdf("FraminghamGeneralElectionResults2017.pdf", format = "xlsx-single", api_key = "YourAPIKeyHere").`

Or, you can use the already-converted file *data/FraminghamGeneralElectionResults2017.xlsx* in this book's GitHub repository.

Next, we want to import the City Council results from the spreadsheet – ideally without all the other races. A basic `rio::import("myfilename.xlsx")` command would pull in *all* the data from that spreadsheet. Fortunately, there's a way to indicate exactly which cells you want, thanks to the readxl package.

Do we need to use readxl::read_xlsx instead of rio::import? It turns out they're the same function.

rio's import() can be thought of as a wrapper, or alias, for other packages' functions. The beauty of rio is that you don't have to remember which package and function to use for different types of data files. So, rio::import() calls readxl::read_xlsx() automatically on an .xls or .xlsx file, without you having to remember which package to use. However, sometimes it's helpful to know specific capabilities of the underlying packages and functions.

You can see which packages rio uses for each file format by running `help("import", package = "rio")` in your console. For Excel files, rio is using the readxl package's read_xlsx() function. And, readxl can import *a specific range of cells*, not just an entire worksheet, with the syntax `readxl::read_xlsx("myspreadsheet.xlsx", sheet = 1, range = "A1:K29")` (or whatever sheet number and range of cells you want).

What cells *do* we want? Check the spreadsheet. The 4th row has headers, rows 11 through 16 and columns A through S hold the data (we don't need column T with the totals).

There are a few different strategies for this. One is to import only rows 11 to 16 and then manually add column names. Another is to import rows 4 through 16 and then delete the first 6 rows of imported data (which contain mayoral data and the unhelpful "At Large City Council (2 seats) for 2 years" header).

Bracket notation works just fine for deleting rows by number, using the syntax `mydf[-c(RowsToDelete),]`

```
myspreadsheet <- "data/FraminghamGeneralElectionResults2017.xlsx"
mycells <- "A4:S16"
council <- readxl::read_xlsx(myspreadsheet, range = mycells)
council <- council[-c(1:6),]
```

Note that those last two lines can be combined as `council <- readxl::read_xlsx(myspreadsheet, range = mycells)[-c(1:6),]`. As is often the case, the tradeoff is whether you want code that's more compact or more readable for beginners.

Let's take a look at the data (Figure 14.3):

```
head(council)
```

Those are rather ghastly column names. **X__1** was a default for the Candidate column because the spreadsheet had no header for that column; and the precinct column names are, um, interesting. It's easy enough to change those with `names()`. Instead of tediously writing out c("Candidate", "1", "2", "3" ...) etc., you can get the character equivalent of 1:18 with `as.character(1:18)`. (I'm turning those into characters because R column names shouldn't start with a number.)

```
## # A tibble: 6 x 19
##   X__1     `1.00000000` `2.00000000` `3.00000000` `4.00000000` `5.00000000`
##   <chr>           <dbl>        <dbl>        <dbl>        <dbl>        <dbl>
## 1 Blanks           642.         660.         464.         716.         513.
## 2 George~          638.         599.         485.         611.         599.
## 3 Pablo ~           89.         104.         107.         103.          95.
## 4 Christ~          645.         808.         467.         583.         594.
## 5 Cheryl~          724.         775.         520.         917.         660.
## 6 Write-~            8.          10.          11.           8.           3.
## # ... with 13 more variables: `6.00000000` <dbl>, `7.00000000` <dbl>,
## #   `8.00000000` <dbl>, `9.00000000` <dbl>, `10.00000000` <dbl>,
## #   `11.00000000` <dbl>, `12.00000000` <dbl>, `13.00000000` <dbl>,
## #   `14.00000000` <dbl>, `15.00000000` <dbl>, `16.00000000` <dbl>,
## #   `17.00000000` <dbl>, `18.00000000` <dbl>
```

Figure 14.3: Data after initial import from the election results spreadsheet.

```
names(council) <- c("Candidate", as.character(1:18))
head(council)
```

```
## # A tibble: 6 x 19
##   Candidate      `1`   `2`   `3`   `4`   `5`   `6`   `7`   `8`   `9`  `10`
##   <chr>        <dbl> <dbl> <dbl> <dbl> <dbl> <dbl> <dbl> <dbl> <dbl> <dbl>
## 1 Blanks        642.  660.  464.  716.  513.  604.  543.  591.  310.  181.
## 2 George P. K~  638.  599.  485.  611.  599.  620.  499.  437.  299.  162.
## 3 Pablo Maia     89.  104.  107.  103.   95.   93.   90.  106.   93.   37.
## 4 Christine A~  645.  808.  467.  583.  594.  599.  379.  459.  248.  167.
## 5 Cheryl Tull~  724.  775.  520.  917.  660.  686.  687.  575.  320.  224.
## 6 Write-Ins       8.   10.   11.    8.    3.    8.    6.   16.    4.    3.
## # ... with 8 more variables: `11` <dbl>, `12` <dbl>, `13` <dbl>,
## #   `14` <dbl>, `15` <dbl>, `16` <dbl>, `17` <dbl>, `18` <dbl>
```

That's better.

Note: If you know in advance what you want your column headers to be, you can include them in the initial import with a format like `council <- readxl::read_xlsx(myspreadsheet, range = mycells, col_names = c("Candidate", as.character(1:18)))`.

Why did I use `readxl::read_xlsx()` instead of `rio::import()` here? Using the actual function instead of the rio::import() wrapper can be useful if you want to take advantage of *RStudio autocomplete for function arguments.* RStudio knows if you start typing `col_n` inside `read_xlsx()` that you want the `col_names` argument. It *won't* know that for the `rio::import()` wrapper.

14.7 Tidying the data

Next, I'd like a "long" data frame with columns for Candidate, Precinct, and number of votes, such as:

```
##             Candidate Precinct Votes
## 9          Pablo Maia        2   104
## 10  Christine A. Long        2   808
## 11 Cheryl Tully Stoll        2   775
## 12          Write-Ins        2    10
## 17 Cheryl Tully Stoll        3   520
## 18          Write-Ins        3    11
```

etc. In other words, tidy data.

There are several different packages that make it easy to reshape a data frame from "wide" – with important information embedded in column names – to "long". I recently transitioned to the tidyverse's tidyr package, although I was a long-time holdout using Hadley Wickham's older reshape2 package.

To create a "narrow"" tidy data frame with tidyr, use the gather() function. I think of it as gathering up all the column information that's spread out over multiple column names but shouldn't be, and putting them into a single new column. And in fact, that's what we want to do here: "gather" all the precinct information that's currently scattered across 18 different columns into a single column named "Precinct."

Here's the compact version: `gather(mydf, newVariableColumnName, newValueColumnName, ColumnsYouWantToGather)`. If you want to make it more understandable by adding the gather() function's argument names as well as argument values, it would be `gather(data = mydf, key = newVariableColumnName, value = newValueColumnName, ColumnsYouWantToGather)`. The "key" is the name of the new column holding category names, the "value" is the name of the new column holding values, and the remaining arguments are the columns you would like to gather.

For the council data, I'd like the new variable column to be named Precinct, the new value column name to be Votes, and all the columns to be gathered except Candidate:

```
council_long <- gather(council, Precinct, Votes, -Candidate)
head(council_long)
```

```
## # A tibble: 6 x 3
##   Candidate            Precinct Votes
##   <chr>                <chr>    <dbl>
## 1 Blanks               1         642.
## 2 George P. King, Jr.  1         638.
## 3 Pablo Maia           1          89.
## 4 Christine A. Long    1         645.
## 5 Cheryl Tully Stoll   1         724.
## 6 Write-Ins            1           8.
```

Notice that the column names don't need to be in quotation marks, and you can also specify just the column(s) you *don't* want to gather by using the minus sign before the name(s).

With reshape2, you'd use the `melt()` function to create a tidy data frame, and then `dcast()` if you need to go from long back to wide (similar to a pivot table in Excel). I found that naming convention easy to remember – melt and then cast – but the argument syntax within the functions could get a bit cumbersome. That's why I'm sticking with the tidyr package here.

14.8 Reshaping the data

The gathered ("melted") data frame is easier to work with in order to get what we want: showing how each candidate finished in each precinct, and listing the top two candidates per precinct.

dplyr has a few functions that can help with ranking rows in data frames. `top_n(myData, myNumber, myNumberColumn)` will pick the top `myNumber` rows in a data frame based on values in column `myNumberColumn`.

However, we need to worry about ties – picking just the top 2 rows when rows 2 and 3 are tied would leave out some important information. `min_rank()` is a better function for elections, because if two candidates tie for the second spot, they'll both be given a rank of 2, and then the next one down will be ranked fourth.

min_rank()'s default gives the #1 spot to the *smallest* value. To make the *largest* value number one, use the format `min_rank(mydf, desc(mycolumn))`, where `mycolumn` is the column holding values that will be used for ranking.

With those functions in our toolkit, we can find the top two finishers. We want to:

1) Filter out rows with Blanks or Write-Ins.
2) Group by precinct, since we're interested in ranking candidates within precincts.
3) Rank each candidate's finish by precinct.
4) Find the first- and second-place finishers in each precinct.
5) Put this back in a more human-friendly format, with each candidate in his or her own column.

Here's code for the first three tasks and a portion of task 4. This finds the *top* finisher in each precinct, but not yet number-two (you'll see why soon):

```
council_winners <- council_long %>%
  filter(Candidate != "Blanks", Candidate != "Write-Ins") %>%
  group_by(Precinct) %>%
  mutate(
    Precinct_Rank = min_rank(desc(Votes)),
    Top_Finisher = Candidate[Precinct_Rank == 1]
  )
```

Do you understand what this code is doing? The first line creates a new data frame `council_winners` from `council_long`. The filter() command removes rows with Blanks and Write-Ins, so only candidates who were on the ballot remain. The next line groups by Precinct, so subsequent operations take place *within each group/Precinct.* New columns `Precinct_Rank` and `Top_Finisher` are created by mutate().

`min_rank(desc(Votes))` says "Rank the rows in this new data frame by the Votes column, largest values first. And because we're using min_rank(), if two values are tied, rank the tied values the same and then skip the next ranking number" (so if two candidates are tied for the top spot, the next-highest candidate would be ranked third, not second).

The line of code creating the Top_Finisher column uses bracket notation for subsetting within dplyr. Let's take a look at how it works.

With "regular" bracket notation, finding all the candidates where Precinct_Rank equals 1 needs the format `council_winners$Candidate[council_winners$Precinct_Rank == 1]`. That's a bit tedious, though, re-typing *council_winners$* a couple of times. Within dplyr piped commands, you don't need to write out the name of the data frame each time: `Candidate[Precinct_Rank == 1]` is enough. If you add a Second_Place column to council_winners

```
council_winners <- council_long %>%
  filter(Candidate != "Blanks", Candidate != "Write-Ins") %>%
  group_by(Precinct) %>%
  mutate(
    Precinct_Rank = min_rank(desc(Votes)),
    Top_Finisher = Candidate[Precinct_Rank == 1],
    Second_Place = Candidate[Precinct_Rank == 2]
  )
```

You should see an error message `Column`Second_Place `must be length 4 (the group size) or one, not` 2. Why did the top rank work but second place didn't? Try running *only* code to find all candidates ranked number 2, separate from the existing election data frame:

```
council_winners$Candidate[council_winners$Precinct_Rank == 2]
```

```
##  [1] "Christine A. Long"  "Cheryl Tully Stoll" "George P. King, Jr."
##  [4] "George P. King, Jr." "George P. King, Jr." "George P. King, Jr."
##  [7] "George P. King, Jr." "Christine A. Long"  "George P. King, Jr."
## [10] "Christine A. Long"  "George P. King, Jr." "George P. King, Jr."
```

```
## [13] "George P. King, Jr." "Christine A. Long"   "George P. King, Jr."
## [16] "Cheryl Tully Stoll"  "George P. King, Jr." "George P. King, Jr."
## [19] "George P. King, Jr."
```

That code works, but do you see the problem? There are *19* second-place finishers, even though there are only *18* precincts in Framingham. In other words, one precinct has a tie for second place. R can't stuff two separate values into a single data frame cell without special instructions, though. That's why the code creating a new column with this data throws an error: You can't add a column with 19 items to a data frame with only 18 rows.

You can see which precinct has the tie by using the table() command on council_winners$Precinct wherever the Precinct_Rank is 2. In the code below, I created a data frame with ranks equaling 2 by using dplyr's filter(), and then ran table() on the result:

```
rank2 <- filter(council_winners, Precinct_Rank == 2)
table(rank2$Precinct)
```

```
##
##  1 10 11 12 13 14 15 16 17 18  2  3  4  5  6  7  8  9
##  1  1  1  1  1  1  2  1  1  1  1  1  1  1  1  1  1  1
```

That's easy enough to eyeball with 18 precincts. If you had many more precincts, though, you might want to just see the precincts with ties.

I like to use the dplyr pipe workflow for cases like this, even though it seems like a lot of code to answer a simple question. What I find so compelling about dplyr is that it encourages you to stop and think, "What do I want to accomplish? What are the steps I need to do so?"

Here, I want to

1) Start with the council_winners data frame,
2) Select only the rows showing second-place finishers – which means filtering for Precinct_Rank equals 2,
3) Count how many times each Precinct shows up, and
4) Select only those Precincts where number of times it appears is greater than 1.

We already know how to do steps 1 (not much of a step) and 2 using `filter(Precinct_Rank == 2)`. For step 3, count() will count the number of times each value in a column shows up, with the frequency column named `n`. Then, we can filter again for any rows where n is more than 1. Here's the code:

```
council_winners %>%
  filter(Precinct_Rank == 2) %>%
  count(Precinct) %>%
  filter(n > 1)
```

```
## # A tibble: 1 x 2
## # Groups:   Precinct [1]
##   Precinct     n
##   <chr>    <int>
## 1 15           2
```

The next question: How can we turn the tied candidate names into a single character string so the overall vector has 18 items, not 19? It might seem that paste() with a separator of " , " between each item should work:

```
myresults <- c("Candidate 1", "Candidate2")
paste(myresults, sep = ", ")
```

```
## [1] "Candidate 1" "Candidate2"
```

Except it doesn't. That still produces a vector with two items, not a single character string that's one item.

This confused me when I was a beginner. `paste("character string one", "character string two", sep = ", ")` produces a single character string from two character strings. But `paste("my vector with 2 character strings")` operating on a *vector* with two strings does not.

Fortunately, paste() has an optional argument, `collapse`, that you can use instead of `sep` to turn a *vector* of character strings into a single string:

```
paste(myresults, collapse = ", ")
```

```
## [1] "Candidate 1, Candidate2"
```

We can use this with each value in the runners-up column to *make sure it's collapsed into a single string.* We'll want to do that with the winners column, too, since we may not know in advance that there are no ties.

So let's modify the code above to make sure all the results are a single character string. Instead of

```
mutate(
  Precinct_Rank = min_rank(desc(Votes)),
  Top_Finisher = Candidate[Precinct_Rank == 1],
  Second_Place = Candidate[Precinct_Rank == 2]
)
```

We want

```
mutate(
  Precinct_Rank = min_rank(desc(Votes)),
  Top_Finisher = paste(Candidate[Precinct_Rank == 1], collapse = ", "),
  Second_Place = paste(Candidate[Precinct_Rank == 2], collapse = ", ")
)
```

Are you still with me? I realize it can be frustrating to spend so much time and thought on such a "little" thing. Unfortunately, that's not uncommon in programming. Remember, though: The good news is, once you've coded this, you can save it and use the script again and again for data in the same format. If you've got results in a similar format where only the column names have changed, it's a single `names(mydf)` command to change column names to what you need. Think of this as an investment. If you create the code once, you can then re-use it – with perhaps slight variations – many times.

14.9 'Long' data back to 'wide'

The data will be easier to analyze in a tidy/long format. But it's easier to scan if you create a new data frame that's in a different but "wide," human-readable format.

Step one: Decide exactly what format you want to present or publish.

A table with *precincts as rows* and *candidates as columns* to display voting results makes sense. Perhaps a second table showing precincts as rows and which candidate won and finished second could also be useful, but first things first.

tidyr's spread() function moves variables from a single column into multiple columns, with each value – in this case candidate – having its own column.

The syntax is `spread(mydf, columnToSpread, columnWithValues)`. So, in this case, we'd want

```
council_table <- spread(council_long, Candidate, Votes)
```

spread() syntax including argument names is `spread(data = mydf, key = columnToSpread, value = columnWithValues)`. Here, the key argument is the column name to spread.

Next, you might want to add a total row to the results. The janitor package has an `adorn_totals()` function that will add a total row, column, or both to a data frame.

In the code below, I'm 1) loading the janitor package (in case you didn't do that at the start of the chapter), 2) replacing council_table with a version of council_table that includes column totals, and 3) looking at the last three rows of council_table with tail() and the n=3 argument.

```
library(janitor)
council_table_with_totals <- adorn_totals(council_table)
tail(council_table_with_totals, n =3)
```

```
## # A tibble: 3 x 7
##   Precinct Blanks `Cheryl Tully Sto~ `Christine A. Lon~ `George P. King, ~
##   <chr>     <dbl>              <dbl>              <dbl>              <dbl>
## 1 8          591.               575.               459.               437.
## 2 9          310.               320.               248.               299.
## 3 Total     7679.              8393.              6844.              6865.
## # ... with 2 more variables: `Pablo Maia` <dbl>, `Write-Ins` <dbl>
```

adorn_totals() defaults to adding a total row, but you can add both total row and column with `adorn_totals(mydf, where = c("row", "col"))`.

If you're creating an HTML table with the DT package, you can use `formatCurrency(colnum, digits = 0, currency = "")` to display column data with commas and still keep the underlying data as numbers. `colnum` would be the column number.

14.10 Winners and runners-up

Last quick task: A table of winners and runners up in each precinct – the council_winners data frame, but in human-readable format. A quick-and-dirty way to do this would be to select just the Precinct, Top_Finisher, and Second_Place columns, but that repeats each precinct 4 times:

```
by_precinct <- council_winners %>%
  select(Precinct, Top_Finisher, Second_Place)
head(by_precinct)
```

```
## # A tibble: 6 x 3
## # Groups:   Precinct [2]
##   Precinct Top_Finisher       Second_Place
##   <chr>    <chr>              <chr>
## 1 1        Cheryl Tully Stoll Christine A. Long
## 2 1        Cheryl Tully Stoll Christine A. Long
## 3 1        Cheryl Tully Stoll Christine A. Long
## 4 1        Cheryl Tully Stoll Christine A. Long
## 5 2        Christine A. Long  Cheryl Tully Stoll
## 6 2        Christine A. Long  Cheryl Tully Stoll
```

As dicussed in Chapter 7, base R's `unique()` function will take care of that, returning a data frame without duplicate rows:

```
by_precinct <- council_winners %>%
  select(Precinct, Top_Finisher, Second_Place) %>%
  unique()
head(by_precinct)
```

```
## # A tibble: 6 x 3
## # Groups:   Precinct [6]
##   Precinct Top_Finisher       Second_Place
##   <chr>    <chr>              <chr>
```

```
## 1 1        Cheryl Tully Stoll Christine A. Long
## 2 2        Christine A. Long  Cheryl Tully Stoll
## 3 3        Cheryl Tully Stoll George P. King, Jr.
## 4 4        Cheryl Tully Stoll George P. King, Jr.
## 5 5        Cheryl Tully Stoll George P. King, Jr.
## 6 6        Cheryl Tully Stoll George P. King, Jr.
```

14.11 Wrap-up

We covered tools for some of the most tedious tasks in data analysis: whipping messy data into shape. We looked at two ways to extract table data out of PDFs: the pdftables R package and (paid) cloud service, or CometDocs. We delved into packages and functions behind rio's simple import() and export() wrappers, including some useful options for reading Excel files with readxl's read_xlsx().

We reshaped data from wide to long and wide again using tidyr's gather() and spread() functions. I also mentioned the reshape2 package's melt() and dcast() not covered in detail here that do similar work.

For elections, we used dplyr's min_rank() function to find winners and runners-up, and to account for ties. Other useful functions include:

- janitor's tabyl for tallying counts of items in a vector (including percents) and adorn_totals() for adding a total row and/or column to a data frame;
- paste() with the collapse argument to turn a multi-item vector into a single character string.

Next up: Combine text, graphics, and scripts to create a single HTML file or Word doc with R Markdown.

14.12 Additional resources

RStudio has a webinar on data wrangling at:

https://www.rstudio.com/resources/webinars/data-wrangling-with-r-and-rstudio/.

I published a video screencast with article and code, Reshape data in R with the tidyr package, at https://bit.ly/reshapetidyr.

14.13 Using tabulizer to unlock the City Council data

Here's the code I'd use for the tabulizer package and the Framingham election data:

```
pacman::p_load(tabulizer)
myfile <- "data/FraminghamGeneralElectionResults2017.pdf"
mydf_list <- tabulizer::extract_tables(myfile, output = "data.frame")
page1 <- mydf_list[[1]] %>%
  slice(-c(1:13)) %>%
  select(-contains("."))
page2 <- mydf_list[[2]] %>%
  slice(1:8) %>%
  select(-contains(".")) %>%
  mutate(
    X1 = vector(mode = "integer", length = 8),
    X2 = vector(mode = "integer", length = 8)
```

```
  )
myresults <- rbind(page1, page2)
```

The first line installs or loads the tabulizer package. Line 2 sets the location of the PDF.

Line 3 extracts all tables tabulizer finds in the PDF and stores them in a list called mydf_list. I've also specified I want the output to be a data frame. tabulizer's default is a matrix, which means every column must be the same type. And since the first column in our tables is character strings of candidate names, all the other columns would come in as characters, too.

I create a data frame from the page 1 table called page1, removing the first 13 rows as well as any column that contains a `.` in the name. dplyr's select includes several handy functions for identifying column names by patterns, such as starts_with and matches as well as contains. See https://dplyr.tidyverse.org/reference/select.html#useful-functions for more.

I create a data frame from the page 2 table called page2, selecting just the first 8 rows and removing the columns where the name contains a `.`. When I do that, I see that the page2 data frame is missing X1 and X2 columns, which will be needed if I want to create one data frame from the two. So I add X1 and X2 columns as vectors with no available data.

In the last line, I create one data frame called myresults from the two, binding them together by rows (with page1 at the top and page 2 at the bottom), requiring both have the same structure and column names.

Chapter 15

Integrate R With Your Storytelling Using R Markdown

Project: Mixing text and R code about that snow data

15.1 What we'll cover

- Creating a basic R Markdown document
- Using the same R Markdown template and variables to generate multiple HTML files

15.2 Packages needed in this chapter

You'll need to install and load the `rmarkdown` package from CRAN. It should install the `knitr` package as well.

```
pacman::p_load(rmarkdown, dplyr, fst, scales, purrr, rio)
```

15.3 R Markdown basics

How do you create a story or report that includes data? One all-too-common workflow is writing text in one place, such as Word or Google Docs; generating calculations, charts, and visualizations in another piece of software, such as Excel or now R; and then doing a lot of cutting and pasting.

With R Markdown, though, *you can do everything in one place:* write, edit, and spellcheck your narrative; write and run R code; and add R-generated summaries, tables, and visualizations.

This offers more than just convenience. An R Markdown document is an elegant way to display, run, and explain your R code if you want readers to see how you came up with your results. It also makes your work reproducible by others – an important standard in scientific research that is becoming increasingly important in data journalism, too. *You* can also make use of reproducible work, if you want to do similar work with new data.

Not all audiences want to see your code, of course. With R Markdown, though, you can add and run R code in the document but hide the code itself, so readers only see the results of that code. This is extremely useful for publishing stories with data and graphics, since an R Markdown document can be used to generate HTML, Word, PDF, and several other types of files.

R Markdown is one of my favorite things about modern R. I use it when working on all but the simplest projects to document my work step by step. I've used it for personal projects, such as whipping up a quick map and chart of local elections results along with a few paragraphs of analysis. I've set up an R Markdown document to read a single spreadsheet of election results and then update multiple tables and graphs that can be easily uploaded to the Web. And it's great when I want to share my work with others.

I also wrote this entire book in R Markdown, using an additional package called bookdown. And I maintain a hyperlocal blog using R Markdown and a package called blogdown.

15.4 Create an R Markdown document

You can create an R Markdown file in RStudio by going to File > New File > R Markdown.

R Markdown files are slightly different from the R script files we've been working with so far. They:

- Have the file extension `.Rmd`
- Must start with a particular header format
- Use a specific syntax when incorporating R code, and
- Use an simpler-than-raw-HTML syntax for text.

This will be easier to understand with an example. Let's first create a simple R Markdown document for the Boston snowfall data in Chapter 6.

In RStudio, go to `File > New File > R Markdown`. Enter a document title, accept the default output format as HTML, and hit enter. RStudio will generate the shell of an R Markdown document. The top of the document will look something like:

```
---
title: "Boston Snowfall"
author: "Your Name"
date: "November 20, 2018"
output: html_document
---
```

That *metadata* is in a format called YAML. (I used to think that stood for Yet Another Markup Language, but in fact is an abbreviation for YAML Ain't Markup Language). You can find out more than you're ever likely to want to know about YAML at yaml.org. But for R Markdown beginners, there are two main YAML-related points to keep in mind: 1) Putting things on their proper lines matters, and 2) category names and their values are separated by a colon, such as `title: "My Election Analysis"` or `output: html_document`.

After that YAML metadata header comes the document body. Some of it can be text, and some can be R code.

15.5 R Markdown text syntax

R Markdown is designed to be a simpler way of formatting text. If you're familiar with HTML coding, you know that <p> starts a paragraph and </p> ends it; <i></i> will put text in italics and <h1></h1> tags are for large headlines. There are lots of other HTML and CSS tags.

````
```{r setup, include=FALSE}
knitr::opts_chunk$set(echo = TRUE)
```
````

Figure 15.1: An R code chunk within an R Markdown document.

Basic HTML isn't all that onerous, but it's nice not to have to worry about paragraph tags. Two carriage returns are all you need to create a paragraph. One underscore or asterisk before and after text _*makes it italics*_. two underscores or asterisks ****make it bold****.

And, instead of <h1>, <h2>, and other h tags for headlines and subheads, you can use the `#` symbol: One `#` for the largest <h1> tag, two for <h2> and so on. In this case, you only need the # or #s at the start of text, not also at the end, but *it has to be at the very beginning of a line.* Putting `#` after a space on its own line won't generate headline text.

There's also a markdown "shortcut" for html links. Instead of having to type out <a href=http://myurl.com>link text</a> for a hyperlink, you can use [link text](http://myurl.com).

You *can* use conventional HTML in the text portion of an R Markdown document as well. And if you know CSS, you can add a CSS stylesheet to an R Markdown document with

```
output:
  html_document:
    css: mystylesheet.css
```

in the YAML header, where "mystylesheet.css" is a separate CSS file in the same directory as the R Markdown file.

For this chapter's sample R Markdown document, write a few sentences about the snowfall data as if for a news story or blog post. Or, just write some dummy placeholder text. You can import the Boston snowfall data with the code below if you'd like to take another look at it:

```
bostonsnow <- rio::import("data/BostonChicagoNYCSnowfalls.csv") %>%
  select(Winter, Boston)
```

Above RStudio's top left script panel, you can see a spellcheck icon (with ABC and a check mark next to the disk Save icon) you can use to check spelling in your text.

Once you've got your sample text, it's time to add some R.

15.6 R code chunks

In the body of a default R Markdown document generated by RStudio, you'll see a few things that may not look familiar.

Figure 15.1 is an *R code chunk:* R code that runs *from within your text document.* R code chunks need to start with three back ticks and `{r}`. And somewhat like functions, R code chunks can have names and some additional options, with a format such as `{r mychunkname, option1=something, option2 = somethingelse, ....}`

`mychunkname` can be anything you name your chunk, as long as each chunk in the document has a unique name. *Make sure you don't repeat chunk names.* Your document won't execute properly if a chunk name is used more than once. Options tell R how to process the code chunk.

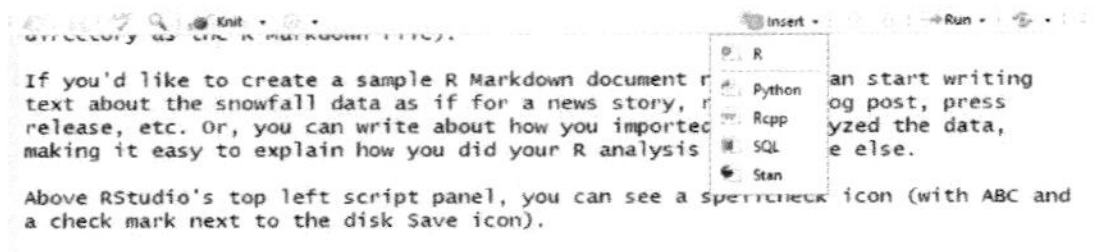

Figure 15.2: RStudio menu for inserting a R code chunk into an R Markdown document.

Figure 15.3: A green arrow lets you execute a code chunk from within the document.

15.7 Adding R code to run

The easiest way to insert R code in the document is to use the Insert dropdown menu just above and to the right of the source code pane (see Figure 15.2).

You'll see choices for several different types of code; pick the first one for R.

When the code to create an R chunk appears, you'll probably notice a few "extras" at the top right of the chunk.

That green arrow at the far right (see Figure 15.3) lets you run code *just within that chunk.* The down triangle with the green line under it is to run any and all chunks *above* the current chunk. There are additional options in the Run drop-down list just above and to the right of the source pane, including a command to run all chunks in the document.

Finally, that barely visible gear icon at the top right of the chunk can help you add or change the chunk's options, if you don't know all the free-form syntax.

Some important chunk options and what they tell R to do:

- `echo = FALSE` – This says, "Don't display the R code in the final 'knitted' document." Knitting is a phrase used to generate an HTML, Word or other file from an R Markdown document. echo = FALSE is useful when you want to do things like generate visualizations for a Web post or Word document for an audience that wouldn't care about (or understand) your code. The results display; the code does not. The default is echo = TRUE. To use this or any other option, include it in the `{r}` brackets, such as `{r echo = FALSE}` .
- `eval = FALSE` – "Don't actually run the code, just show it." Useful if you're writing an R tutorial where you want to show sample code without executing it. That also defaults to TRUE.
- `warning = FALSE` and `message = FALSE` – "Don't show warnings or messages when executing code."
- `comment = ""` – R Markdown documents default to having `##` at the beginning of each line of code results. This is helpful if you're creating a document that displays your code and its results, but not so useful if you just want to show a general audience your results without the code. Setting comment to "" in a chunk changes those default double-# characters to nothing.

If you're adding graphics to your chunk, you can set options for width and height in inches with `{r, fig.width=6, fig.height=4}`.

15.8 Add an R-generated graph

Next, we'll add one of the graphs from Chapter 6 to your sample, for example, the one showing 10 snowiest Boston winters.

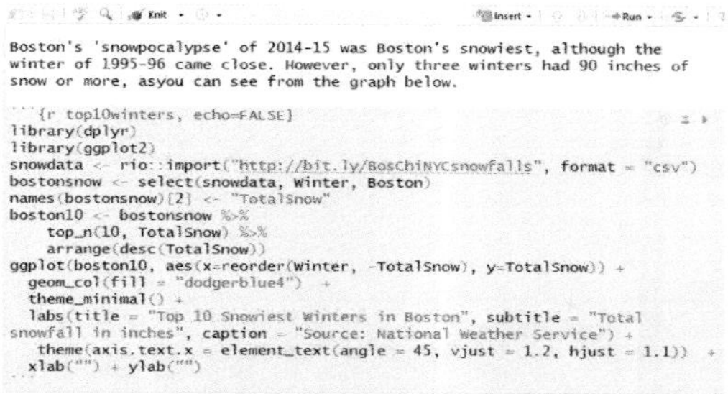

```
Knit   Insert   Run

Boston's 'snowpocalypse' of 2014-15 was Boston's snowiest, although the
winter of 1995-96 came close. However, only three winters had 90 inches of
snow or more, asyou can see from the graph below.

```{r top10winters, echo=FALSE}
library(dplyr)
library(ggplot2)
snowdata <- rio::import("http://bit.ly/BosChiNYCsnowfalls", format = "csv")
bostonsnow <- select(snowdata, Winter, Boston)
names(bostonsnow)[2] <- "TotalSnow"
boston10 <- bostonsnow %>%
    top_n(10, TotalSnow) %>%
    arrange(desc(TotalSnow))
ggplot(boston10, aes(x=reorder(Winter, -TotalSnow), y=TotalSnow)) +
  geom_col(fill = "dodgerblue4")  +
  theme_minimal() +
  labs(title = "Top 10 Snowiest Winters in Boston", subtitle = "Total
snowfall in inches", caption = "Source: National Weather Service") +
   theme(axis.text.x = element_text(angle = 45, vjust = 1.2, hjust = 1.1))  +
  xlab("") + ylab("")
```
```

Figure 15.4: Code chunk to display a graph.

It's important to treat your R Markdown document as a self-contained R environment, separate from your working R session in RStudio. That means you'll need to load your packages and data in the document, even if they're already loaded into your current working environment.

Add something like code in Figure 15.4 to the R Markdown document, then click the Knit icon just above the source pane. You should see a graph similar to Figure 15.5.

You can "knit" an R Markdown document into an HTML file from the console command line or within another R script (so you don't have to keep clicking that Knit button) with

```
rmarkdown::render("myRMarkdownFileName.Rmd",
    output_format="html_document", output_file = "myhtmlfile.html" )
```

As I mentioned earlier, you've got other choices for an output file, including PDF (with appropriate software on your system) and Word. There are also several HTML slideshow options, such as ioslides and beamer. And as of the August 2018 RStudio preview (beta) version, you can create PowerPoint files. Check out the R Markdown website at rmarkdown.rstudio.com for more information on those and other formats.

15.9 Setting options

A few other points on R Markdown options:

In addition to setting an option for each chunk, you can set options for *all* the chunks in a document at the outset, using the command `knitr::opts_chunk$set(myoption = myoptionvalue)` such as in Figure 15.6.

That needs to be inside an R code chunk.

Some things involving HTML that aren't R-specific, such as a default figure size, can be set in the YAML header.

15.10 Mixing R within text

There's another way of displaying *the results* of R code within the markdown document's text. If you have an R code chunk that stores candidate Smith's vote total in a variable called `smithvotes`, you can display the *value* of `smithvotes` with a syntax such as

```
Candidate Smith received `r smithvotes` in the election.
```

You can run actual R code in there as well, such as ``Smith received `r sum(mydf$Smith)` votes.``

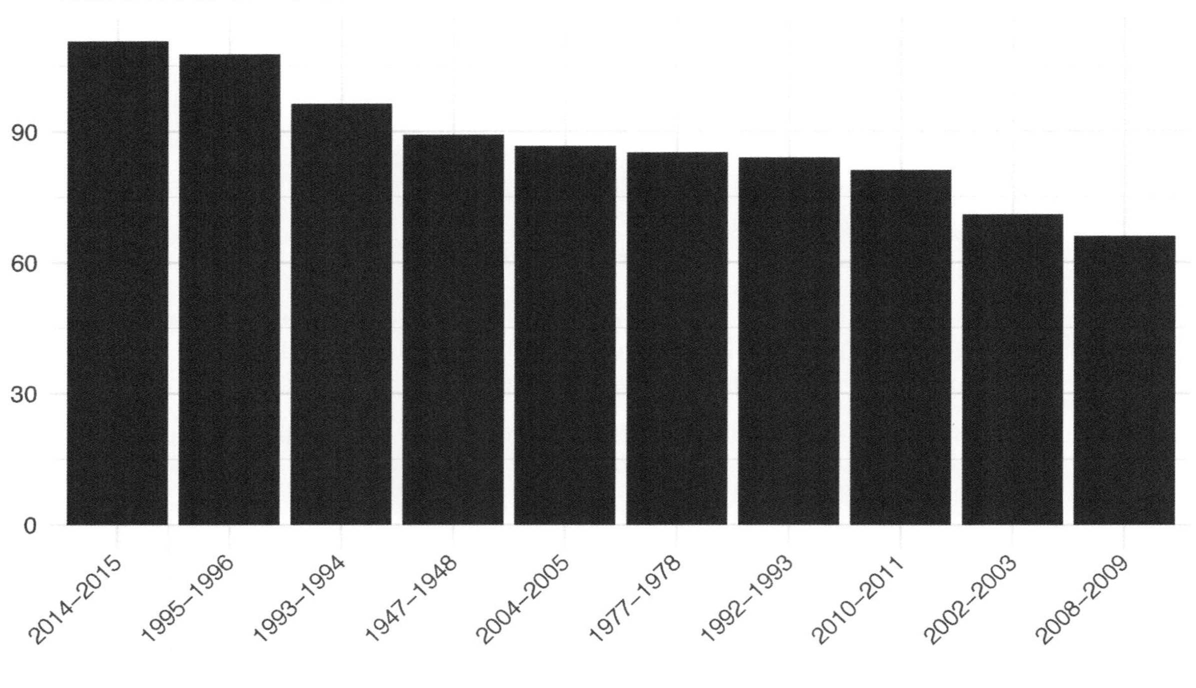

Figure 15.5: Resulting graph from an R Markdown code chunk.

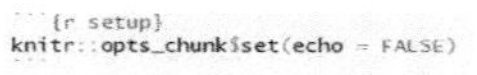

````
```{r setup}
knitr::opts_chunk$set(echo = FALSE)
```
````

Figure 15.6: R Markdown setup options.

15.11 Even more options

15.11.1 Table of contents

You can include a table of contents in your HTML document by adding this to your YAML:

```
---
title: "Your Title"
output:
  html_document:
    toc: true
---
```

That creates a table of contents at the top of your document, If you want to change its location, you can make it "float" – that is, appear on the left if the viewer's browser window is wide enough but on top if it's narrow. Do this with the `toc: true` *and* `toc_float: true` options:

```
output:
  html_document:
    toc: true
```

```
    toc_float: true
```

If you want to keep things simple, though, the only part you *really* need if you want to generate an HTML document is:

```
output: html_document
```

Even the title is optional (although it's usually a good idea to have a title for an HTML document).

Other output choices include `word_document` for a Microsoft Word file, `pdf_document` for PDFs, and `ioslides_presentation` for an HTML slideshow format.

Not surprisingly, you'll need Word on your system to create Word documents, and some additional system installations for PDFs. (Generating a PDF document from R on a Windows machine is often not a trivial exercise, even after installing the necessary TeX software).

15.11.2 Themes and templates

In addition to creating your own look with custom CSS as mentioned earlier, you can also use some pre-made R Markdown themes and templates. The rmarkdown package includes themes from the Bootswatch collection (which includes Bootstrap themes).

As of this writing, available themes included default, cerulean, journal, flatly, readable, spacelab, united, cosmo, lumen, paper, sandstone, simplex, and yeti. All were available for preview at bootswatch.com.

Apply a theme in an R Markdown document's YAML header with

```
output:
  html_document:
    theme: spacelab
```

(or whatever theme you want).

Prolific R coder Bob Rudis has released several R Markdown templates that don't require Bootstrap or jQuery. You can find out more about them at https://github.com/hrbrmstr/markdowntemplates.

15.12 Repeatability with R Markdown parameters

A single R Markdown document is useful for creating a single document or slideshow. But what if you want multiple pages based on data with the same format? For example, you might have demographic data about multiple neighborhoods or towns, and you'd like to generate a separate HTML page for each.

You can do that, too.

R Markdown parameters let you use one R Markdown document to generate multiple HTML or Word docs that are basically the same *except for one or more specific variables.*

For this example, I'll use a data file with information about healthcare expenditures from 2000 to 2015 for 10 countries in Europe and North America. The data come from the World Health Organization at http://apps.who.int/nha/database/Select/Indicators/en. The file, "WHO_Health_Spending_Per_Capita.xlsx", is in the data subdirectory of files from this book's GitHub repo.

I'll read in the spreadsheet and do some basic cleanup such as deleting two unneeded columns, deleting a row with column-name information (column information in this spreadsheet is unhelpfully spread over two rows instead of one), making column names more R-friendly (which means not starting with a number), and turning all the data that was imported as characters back to numbers.

The numbers come in as characters because there's *one* row of data with character strings instead of numbers. I'll delete that first row after the headers, but the damage will be done: Since each data frame column has to be all one type, one single character string in a column will turn all the other numbers into character strings as well.

While I'm a big fan of dplyr and the tidyverse, I'd now like to show you base R's apply() function as a way to apply a function to values in a data frame, either by column or by row. It's a handy thing to know at times.

The syntax is `apply(mydata, margin, myfunction)` where margin is a number indicating whether myfunction should be applied by *row* or *columns*. When margin is 1, the function is applied by row; if it's 2, the function is applied by column.

The code below deletes a row with some header information, removes unneeded columns, and looks at the first three rows of the resulting data frame:

```
healthdata <- rio::import("data/WHO_Health_Spending_Per_Capita.xlsx")
# Get rid of row 1 and columns 2 and 3
healthdata <- healthdata[-1,-c(2:3)]
# Change all the numeric data that imported as characters -- in all the columns
# except column 1 -- back to numbers.

healthdata[,-1] <- apply(healthdata[,-1],2,as.numeric)
head(healthdata, 3)
```

```
##                  Countries       2000       2001       2002       2003
## 2                   Canada 1345.78209 1326.48536 1370.58704 1821.14741
## 3                   Mexico   32.74617   37.88166   38.86123   37.20955
## 4 United States of America 4561.91335 4912.70582 5330.04350 5738.91196
##         2004       2005       2006       2007       2008       2009
## 2 2239.31772 2709.19709 3270.56363 3840.79063 4113.49466 3753.69392
## 3   38.56904   43.32105   45.96995   49.66841   50.00439   35.34612
## 4 6101.49082 6453.09599 6820.59911 7176.61553 7420.06732 7696.47151
##         2010       2011       2012       2013       2014      2015
## 2 4841.51409 5348.43468 5347.95470 5133.76996 4546.56866 3524.0072
## 3   42.63568   45.48937   44.09813   48.37732   44.64177   33.7457
## 4 7949.89613 8160.84486 8432.50653 8634.63175 9059.52116 9535.9453
```

That second-to-last line of code's apply() says "for the healthdata data frame: Exclude column 1, apply the as.numeric() function on all the data by column, and store results in the R object healthdata (again excluding the first column)."

If you'd prefer to do this the tidyverse way, dplyr has a function called `mutate_at()` that uses a syntax `mutate_at(vars(my_columns_to_mutate), myfunction)`. The `vars()` needs to be part of the code, and then you can use a logical test to pick the columns. In this case, all the columns I want to change start with "2", and dplyr has a `starts_with()` function. So, this could take the place of `apply(healthdata[,-1],2,as.numeric)`:

```
healthdata <-  mutate_at(healthdata, vars(starts_with("2")), as.numeric)
```

As always, unless you're working with enormous data sets where it's worth comparing the performance of each, this comes down to syntax personal preference.

15.12.1 Generate a report for each country

One obvious analysis to do on this data is to compare data for different countries. I'll leave you to do that using packages like dplyr and ggplot2. However, for this exercise, let's say we *also* want to present a separate report or Web page for each country. R Markdown makes it easy, offering a tool that would be useful for election results by state, demographic data by county, and a lot of other data that you might want to present by category in addition to analyzing the entire set as a whole.

I'll save that data frame as an R object with `fst::write.fst(healthdata, "healthdata.fst")`. I chose the fst package because it is extremely fast to read and write data; but I also could have saved it as an R object with `saveRDS(healthdata, "healthdata.Rdata")` or CSV file with `rio::export(healthdata, "healthdata.csv")`.

First, I'll create an R Markdown document which reads in that healthdata.fst file and graphs U.S. data. It uses tidyr's `gather()` function to create a 2-column data frame of "key-value" pairs. And, to keep the R relatively simple, I'll use a basic, default-theme ggplot2 bar graph – but to make that somewhat interactive for an HTML page, I'll wrap that in plotly's ggplotly() function.

I'd also like to use in-text R code to display the percent change from 2000 to 2015. My goal is to generate text along the lines of "Per-capita spending rose 109% from 2000 to 2015," but I don't want to hard code that, because I'd like this document to be re-usable for other countries.

The percent change would be `(last row value - first row value) / first row value` displayed as a percentage. In R, that would be `(graphdata$Spending[16] - graphdata$Spending[1]) / graphdata$Spending[1]`. To get the fraction displaying as a percent, I can use scales::percent(), such as `scales::percent((graphdata$Spending[16] - graphdata$Spending[1]) / graphdata$Spending[1])`.

Even more generically, instead of hard-coding 16 for the last row, that could be replaced by `nrow(graphdata)`:

```
scales::percent((graphdata$Spending[nrow(graphdata)] - graphdata$Spending[1]) /
graphdata$Spending[1])
```

It would be nice to be able to say that spending *rose* or *fell* by a certain amount, instead of displaying a more general word such as "changed." I usually use an R function like this to generate my verb:

```
getverb <- function(mychange){
  myverb <- case_when(
    mychange > 0 ~ "rose",
    mychange < 0 ~ "fell",
    TRUE ~ "was unchanged"

  )
  return(myverb)
}
```

This uses dplyr's case_when() function to set the value of `myverb` to be "rose" if my data point is greater than 0, "fell" if it's less than 0, and "was unchanged" in other cases. `getverb(.16)` would return "rose"; `getverb(-.004)` would return "fell".

One last point for making a more elegant text display. If spending decreased and the change is a negative number, I don't want the text to read "Spending fell -5.6%". So, when using an R-generated verb like this, I don't want to display the negative, which means I want the text to show the *absolute value* of any change percent.

An R Markdown document incorporating all these points might look something like Figure 15.7.

Now, though, we'd like to replace the hard-coded "United States of America" with an R Markdown variable for country. That's where parameters come in.

R Markdown parameters are defined in a document's YAML header, using the format:

```
params:
  parametername: parameterdefaultvalue
```

So in this case, the document could include this in the header:

````
---
title: ""
output: html_document
---

```{r setup, echo=FALSE}
knitr::opts_chunk$set(echo = FALSE)
suppressPackageStartupMessages(library(ggplot2))
suppressPackageStartupMessages(library(plotly))
suppressPackageStartupMessages(library(dplyr))
country <- "United States of America"
graphdata <- fst::read.fst("healthdata.fst") %>%
  filter(Countries == country) %>%
  select(-Countries) %>%
  tidyr::gather(key = "Year", value = "Spending")

getverb <- function(mychange){
  myverb <- case_when(
    mychange > 0 ~ "rose",
    mychange < 0 ~ "fell",
    TRUE ~ "was unchanged"

  )
  return(myverb)
}

rawchange <- graphdata$Spending[nrow(graphdata)] - graphdata$Spending[1]
```

# `r country` Healthcare Spending Per Capita

Spending `r getverb(rawchange)` `r scales::percent(abs(rawchange / graphdata$Spending[1]))`.

```{r spending_graph, warning=FALSE, message=FALSE}
ggplotly(
ggplot(graphdata, aes(x=Year, y=Spending)) +
  geom_bar(stat="identity")
)
```
````

Figure 15.7: R Markdown document for spending per capita.

```
params:
  country: Canada
```

To use that variable in the document, the format is `params$parametername` – or in this case, `params$country`.

So now I can change `country <- "United States of America"` to `country <- params$country` within the document, and leave the rest of the `country` references intact.

Finally, to generate (knit) an HTML document using a parameter variable, the format for a single document is `rmarkdown::render("Markdown_document_name.Rmd", output_file = "my_html_file_name.html", params = list(parametername = "parametervalue"))`. In other words, to generate an HTML file for Mexico, you could run

```
rmarkdown::render("Health_Expenditures.Rmd", params = list(
  country = "Mexico"
))
```

If you don't specify a file name, the file will use the name of the R Markdown document and append the appropriate html, doc, or other extension.

What we want, though, is to generate a document for each country, with a different file name for each, without having to write out the render command for each country one by one. We can write a simple function that will do this for each country name:

```
get_country_page <- function(country_name, markdown_file = "Health_Expenditures.Rmd"){
  filename <- paste0("health_expenditures_", country_name, ".html")
  rmarkdown::render(markdown_file, output_file = filename, params = list(country =
 country_name))
}
```

Let's take a look at what's happening there. The `get_country_page()` function takes two arguments: country_name, and the name of a markdown_file that defaults to our "Health_Expenditures.Rmd" file.

The first line of the function creates a *name* for the HTML output file. That name is based on the name of the country_name argument.

The second line of code runs the rmarkdown::render() function on the R Markdown document, specifying the output file as the name we just created, and setting the country parameter to the country_name argument.

15.12.2 purrr's walk() function

walk() is related to purrr's map() functions. They have the same basic goal: apply a function to each value in a vector or list. The difference is that map() and its variations return a value, while walk() doesn't. If you're applying a function where you care about the value returned, use map(). But if you're using a function where you just want *something to happen* such as printing or creating an HTML file, use walk().

To repeat: map() is for when you care about a value that's returned; walk() is for when you don't.

The get_country_page() function is only designed to generate an HTML file; it doesn't return a *value* that we'd want to store *in an R object*. So, walk() is the purrr function we'd want to use for generating the country page HTML files.

One walk() syntax format is `walk(mydata, myfunction)`. It says "apply `myfunction` to each value in `mydata`".

Since we'd like to apply the `get_country_page` function to each value in healthdata$Countries, this is one way to use walk():

```
library(purrr)
walk(healthdata$Countries, get_country_page)
```

15.12.3 Another walk() and map() syntax: formulas

There's another format worth knowing about for using walk() and map(), though: with a *formula* instead of a function. With a formula, you don't use a separate function with a name like get_country_page. Instead, all the necessary instructions go right inside walk() or map().

I find writing separate functions is often clearer than working with formulas. Feel free to skip this section for now if it feels like it's getting too complicated. But I do want to introduce the concept.

Hopefully, an example will make it clear. Instead of using a separate get_country_page() function, I'll use a formula – sometimes also called an "anonymous function," since it's basically a function without its own name.

Before writing a formula that can be re-used for multiple cases, it's helpful to see how you'd write the code for just one case.

I often repeat this advice to myself whenever I'm working on even moderately complex code that needs to perform some action on each value in a data set.

So, here's code that generates a `health_expenditures_Canada.html` file for Canada:

```
rmarkdown::render("Health_Expenditures.Rmd",
                  output_file = paste0("health_expenditures_", "Canada", ".html"),
                  params = list(country = "Canada"))
```

How would I turn that into a walk() formula? The syntax for such formulas would be:

```
  walk(healthdata$Countries, ~ rmarkdown::render("Health_Expenditures.Rmd",
                  output_file = paste0("health_expenditures_", .x, ".html"),
                  params = list(country = .x))
  )
```

Let me explain. The walk() code starts with your data set – in this case, the vector of values we want to operate on. That's followed by a comma and a tilde ~ character. The tilde tells walk: "What comes next is a formula." Then, the rest of the code is the same code that worked for one value, Canada – but replacing every instance of `"Canada"` with `.x`. `.x` is a *variable representing the current value in the data set that's being operating on.*

Using a pipe can make this more readable – and when using map() with a return value, the pipe lets you elegantly use results in additional tasks.

```
healthdata$Countries %>%
walk(~ rmarkdown::render("Health_Expenditures.Rmd",
                output_file = paste0("health_expenditures_", .x, ".html"),
                params = list(country = .x))
)
```

When to use a function and when to use a formula? In this case, I prefer the first code with a separate function over cramming the function code within walk(). I think it's easier to understand. But that's pretty much a matter of personal preference. In some cases, a formula might make more sense. It's a good option to have in your R toolkit as your skills get more advanced.

15.13 Wrap-up

We covered how to use R Markdown to combine text, R code for analysis, and R-generated visualizations. In addition, we looked at advanced options such as a table of contents, custom CSS and themes, and parameters to use the same template for multiple sets of data. The chapter ended with a brief introduction to the robust and flexible purrr package.

Next up: Simple Web scraping with rvest, purrr, and robotstxt.

15.14 Additional resources

RStudio's R Markdown Web site at rmarkdown.rstudio.com has a lot of information about creating and customizing documents, notebooks, slide presentations, dashboards with RStudio's flexdashboard package, and more.

For examples of all the things you can do with R Markdown, check out RStudio's R Markdown gallery at http://rmarkdown.rstudio.com/gallery.html.

A video of Garret Grolemund's hour-long session webinar, Getting Started with R Markdown, is available at https://www.rstudio.com/resources/webinars/getting-started-with-r-markdown/.

Looking for some brief ideas of how to tweak an R Markdown doc? Check out Pimp my RMD: a few tips for R Markdown by data analyst Yan Holtz at https://holtzy.github.io/Pimp-my-rmd/.

The University of Edinburgh's Coding Club posted an R Markdown tutorial on GitHub at https://ourcodingclub.github.io/2016/11/24/rmarkdown-1.html.

Chapter 16

Simple Web scraping

Maybe there's information on a Web site that you'd like to analyze. Or perhaps there are a lot of files on a Web site that you want to download – and you don't want to click on each link manually. For a couple of tables, copy and paste might work; for a handful of files, click and save could be fine, too. But for larger and more complex challenges, Web scraping could be the answer.

16.1 Project: Download RStudio PDF cheat sheets

We'll create a simple file-download scraper for all of RStudio's PDF cheat sheets (hopefully their HTML page won't change format between the time this book was published and the time you're reading this!).

16.2 What we'll cover

First, we'll check to make sure the site hasn't blocked automated retrieval. Next, we'll get a list of all the links we want, with the help of a Chrome extension called SelectorGadget and the rvest R package. Finally, we'll download the files.

16.3 Packages needed in this chapter

```
pacman::p_load(rvest, robotstxt, dplyr, purrr)
```

16.4 Step 1: Follow the rules with robotstxt

It's a World Wide Web convention that if a site wants to restrict automated bots from "crawling" its pages – either all pages or just some – it posts those details in a `robots.txt` file in its root directory. So, for a site at www.thesiteurl.com, robots.txt can be found at http://thesiteurl.com/robots.txt (or, in many cases, http://www.thesiteurl.com/robots.txt or https://www.thesiteurl.com/robots.txt).

To be a responsible and considerate Internet citizen, you should make sure a site hasn't refused bots and scripts from accessing its pages before starting to scrape.

You can look at these files manually – for example, checking RStudio's robots.txt file by going to https://wwwrstudio.com/robots.txt in a browser. But it's more elegant – and automated – to use the

Figure 16.1: Download button outlined in orange with SelectorGadget.

robotstxt package to check for you. Plus, each time you run a scraper (if it's one you want to use more than once), you'll be sure the site's robots.txt hasn't changed to exclude you.

To do this, if you haven't run the pacman::p_load() command at the beginning of this chapter, install the robotstxt package with `install.packages("robotstxt")`. We'll also need the rvest package for this project, so install that now as well.

Run `vignette(package = "robotstxt")` and you'll see there's an available vignette called `using_robotstxt`. Run `vignette("using_robotstxt")` to see the instructions. An easy way to use the package, according to the vignette, is with the paths_allowed() function.

At the time I wrote this, RStudio cheat sheets were at https://www.rstudio.com/resources/cheatsheets/ so that's what we'd want to check.

```
library(robotstxt)
paths_allowed("https://www.rstudio.com/resources/cheatsheets/")
```

```
## [1] TRUE
```

Great! The path we want says paths_allowed() TRUE, so we're ready to go.

16.5 Step 2: Get a list of links

If you don't have the Chrome browser installed on your system, do that now, and then go to selectorgadget.com. There should be a link on the page to the SelectorGadget extension. Install it in your Chrome browser, and you should see a new icon for it in the area to the right of your address bar.

It's worth watching the short (less than 2 minutes) video at selectorgadget.com that shows how it works. rvest also includes a vignette explaining SelectorGadget, which you can view with `vignette("selectorgadget")`. In addition, I posted a brief explainer article and video at http://bit.ly/Rscraping.

SelectorGadget is a point-and-click tool that lets you easily figure out what CSS selectors to use in order to extract a portion of a Web page. If you aren't familiar with CSS, you can still use SelectorGadget. You probably already know some simple selectors, such as `p` for a paragraph or `a` for a hyperlink. CSS makes those more useful by categorizing those broad HTML elements into subsets, using identifiers hopefully added by the creator of an HTML page, such as *class* and *id*.

Let's try it out on the RStudio cheat sheet page. Head to https://www.rstudio.com/resources/cheatsheets/ in Chrome, click on the SelectorGadget icon, and then move your cursor over the first download button, making sure the whole button is surrounded by orange (not only the download text – see Figure 16.1), and click on it.

When I wrote this, that single click turned the box around the image green, which means SelectorGadget chose that button. There were also yellow boxes around several more buttons, which means they were selected, too. However, if you scroll down the page, you may see there are other buttons that still aren't selected.

Click on one of those unselected buttons, and you should now see all the buttons chosen. If there were other areas of the page in yellow that you *didn't* want, you could click one of those to de-select it (it would turn red).

When you've got exactly the selection you want, look at the text at the bottom right of the page: That's your CSS selector, in my case `.button-default`, as shown in Figure 16.2.

Copy it and paste it into an R variable.

Figure 16.2: CSS selector shown in SelectorGadget.

```
my_css <- ".button-default"
```

16.5.1 Parse an HTML page

Once you know where your data is, there are several steps to parsing an HTML page with rvest and extracting data.

1. Read in the HTML page, using rvest's `read_html()` function. Load the rvest package first with `library(rvest)` if you didn't do that already with the `pacman::p_load(rvest)` command.

```
library(rvest)
my_html <- read_html("https://www.rstudio.com/resources/cheatsheets/")
```

That will return a list with special R object classes of "xml_document" and "xml_node".

2. Extract the portion of the page we want using CSS selectors and rvest's html_nodes() function. The page we read in is the first argument to html_nodes(); the CSS selector is second.

```
my_nodes <- html_nodes(my_html, my_css)
```

When I wrote this, there were 13 official RStudio cheat sheets plus 10 community-contributed sheets, so my_nodes was a list of 23. Take a look at the first item in the list to see an item's format with

```
my_nodes[[1]]
```

```
## {xml_node}
## <a class="fusion-button button-flat fusion-button-square button-large button-default
## button-2 ButtonCSRM" target="_self" href="https://github.com/rstudio/cheatsheets/raw/
## master/keras.pdf">
## [1] <span class="fusion-button-text">Download</span>
```

If you don't remember why I used double brackets to look at the first item in that list, you may want to refer back to Chapter 7, where I covered subsetting lists and the need for double brackets to inspect the contents of an R list item.

What we want isn't the entire item but *just the link* – the href "attribute" within the tag. Fortunately, there's another rvest function that can extract a specific attribute: the html_attr() function. For the 'href' attribute, `html_attr('href')` would work. Which gives us this extraction code so far:

```
my_urls <- html_nodes(my_html, my_css) %>%
  html_attr('href')
```

Now take a look at the first element of my_nodes.

```
my_urls[1]
```

```
## [1] "https://github.com/rstudio/cheatsheets/raw/master/keras.pdf"
```

That it: We now have a list of links in my_urls.

If instead of the link attribute, you wanted the *text* for that link, you would use the `html_text()` function and not `html_attr()`:

```
my_nodes_text <- html_nodes(my_html, my_css) %>%
  html_text()
my_nodes_text[1]
```

```
## [1] "Download"
```

On this page, all the text is the same: "Download". However, there can be a lot of other cases where the text of the link can be useful.

16.6 Step 3: Download files

Finally, we'd like to download the files at each of those links. We can do that by applying base R's download.file() function. However, download.file() requires both a url (which we have) and a file name (which we don't have yet), using the syntax `download.file(myurl, myfilename)`. purrr can help.

It would be easiest, but pretty clunky, to use the URL as the file name. We'll be happier later if we extract a file name from the URL and use that for the name, though. For example, if we've got a url "https://github.com/rstudio/cheatsheets/raw/master/rmarkdown-2.0.pdf", we'd like just that last part of the URL after the final slash, "rmarkdown-2.0.pdf", to be the file name.

Fortunately, base R has a function to do just that: basename():

```
basename("https://github.com/rstudio/cheatsheets/raw/master/rmarkdown-2.0.pdf")
```

```
## [1] "rmarkdown-2.0.pdf"
```

We want to apply the basename() function to all the URLs in my_urls *and save the results into an R vector of character strings.* purrr's `map_chr()` will do just that.

The code below creates a character vector with file names:

```
my_filenames <- map_chr(my_urls, basename)
```

16.6.1 Apply a function to two vectors at a time

Ideally, I'd like to use the vector of URLs *and* the vector of file names when downloading files, so each URL is downloaded to a file with the appropriate file name. In other words: download.file(my_urls[1], my_filenames[1]), download.file(my_urls[2], my_filenames[2]), and so on.

Both walk() and the map() family have sister functions designed to do just that: apply a function to two data sets at a time, one by one. For walk, it's walk2(). For map, it's map2(), map2_df(), map2_chr(), and so on.

In this case, we want to download the files, but the download.file() function itself saves the file – there's no additional value we want to store. So, a walk() option is the better choice. walk2() uses the syntax `walk2(myfirstvector, mysecondvector, myfunction)` to apply myfunction like `myfunction(myfirstvector[1], mysecondvector[1])`, `myfunction(myfirstvector[2], mysecondvector[2])`, `myfunction(myfirstvector[3], mysecondvector[3])`, and so on. This code applies download.file() to the URLs and file names one by one in tandem, with the URLs as first argument and file names as the second argument:

```
walk2(my_urls, my_filenames, download.file)
```

16.6.2 Why not a for loop?

If you come from another programming language, you'd typically use a `for loop` to solve a problem like this. And you can in fact do this in R, too, with code such as:

```
for(i in seq_along(my_urls)){
  download.file(my_urls[i], my_filenames[i])
}
```

What that code says is: "I'm setting a variable called i to represent each item in a data set. (You can name it anything, but i is commonly used across programming languages for this type of counter). So, for each i starting with 1 through the number of items in my_urls, do download.file(my_urls[i], my_filenames[i])."

In my case, `i` would go from 1 to 23 (there were 23 links to cheat sheets), and the loop would start with `download.file(my_urls[1], my_filenames[1])`, then loop to `download.file(my_urls[2], my_filenames[2])`, and so on.

for loops have long been frowned upon in R, but no less an authority than Hadley Wickham says they're fine to use. In the old days, R was structured in a way that made for loops run very slowly. That's no longer the case – especially if you set up a variable properly in advance for those times when you want a for loop to save results. Wickham believes the purrr-type model is easier to read. I find I tend to make fewer errors, because the code is clearer.

Even if you want to use for loops, it's important to understand how vectorized functions such apply(), walk, and map() work in R, because most of the R code you'll encounter will use them. But if you are more comfortable sticking with for loops for now, there's no deal-breaking reason not to.

Note: If you do decide to go the for-loop route, be advised that `1:length(my_urls)` will work most of the time, but `seq_along(1:length(my_urls))` is the preferred code. Why? `1:length(my_urls)` will generate an unwanted result if my_urls is empty. Try creating an empty vector with `myvec <- c()` and then run `1:length(myvec)`. You'll get a result of 1 and 0. `seq_along(myvec)` returns the proper 0.

16.7 Wrap-Up

First we used the robotstxt package to see whether a Web site permits scraping. Then we used the SelectorGadget Chrome extension to identify portions of a page we wanted to "scrape."

We used the rvest package in order to extract information from a Web page based on SelectorGadget CSS selectors. There were three steps to this process: Read the full Web page with read_html(), extract a portion of the page with html_nodes(), and then home in on exactly what we want from the initial extraction using code such as html_attr('href') for links and html_text() for link text.

Finally, we looked at purrr's walk() and map() functions to apply download.file() to a list of links, including map_chr() to get results as a character vector and walk2() to apply a function to two vectors at once, step by step.

Next up: Analyzing campaign finance data and comparing it with actual election results – all in an R Markdown document.

16.8 Additional resources

The RStudio webinar **"Data Science Case Study"** with Mine Cetinkaya-Rundel includes some Web scraping of the La Quinta Web site to download their locations using SelectorGadget and rvest.

Why La Quinta? To analyze data from the Mitch Hedberg joke that "la Quinta is Spanish for 'next to Denny's'." https://www.rstudio.com/resources/webinars/data-science-case-study/

Charlotte Wickham's purrr tutorials are extremely useful, easy to follow, and packed with helpful tips. A video of her 90-minute presentation at the UseR! conference in Brussels is available at Microsoft's Channel 9 https://channel9.msdn.com/Events/useR-international-R-User-conferences/useR-International-R-User-2017-Conference/Solving-iteration-problems-with-purrr (if you've got the paper version of this book, it might be easier to simply go to channel9.msdn.com and search for Solving iteration problems with purrr.)

An earlier version from the 2017 RStudio conference, without video but with step-by-step slides and code, is on GitHub at https://github.com/cwickham/purrr-tutorial.

If you are a more advanced programmer and interested in using APIs with R, check out the httr package and httr quickstart guide vignette. In addition, Steph Locke has a short how-to at https://itsalocke.com/blog/r-quick-tip-microsoft-cognitive-services-text-analytics-api/.

Chapter 17

An R project from start to finish

17.1 Project: Local political contribution and election data

When voters in Framingham, Massachusetts went to the polls in September 2017, it was a notable preliminary election: The community was shedding its town form of government to become a city. There had never been a mayor before, making it unusually challenging to predict how the seven candidates would finish (top 2 earned a spot on the November ballot).

I thought that data about political contributions could help – but *not* total amount raised. Instead, I wanted to see whether *number of contributors solely within Framingham itself* would help predict the September results.

My reasoning: 1) After raising enough funds for a decent campaign, extra money wouldn't help all that much (it's not like candidates were buying pricey TV ads); and 2) a preliminary election attracts the most committed and engaged voters, who are most likely to give even a few dollars to the candidate of their choice.

Naturally, I used R every step of the way. Here's how.

17.2 What we'll cover

- Standardizing names to join two data frames
- The briefest of looks at regular expressions
- Calculating correlations
- Creating an interactive scatterplot
- Visualizing a linear regression

17.3 Packages needed in this chapter

```
pacman::p_load(purrr, dplyr, stringr, janitor, magrittr, ggplot2, here, readr, tidyr)
pacman::p_load_gh("hrbrmstr/taucharts")
```

17.4 Get the data, make it ready for analysis

First, I needed to download all candidates' initial campaign finance reports from the Massachusetts Office of Campaign and Political Finance. You can find those files in the book's GitHub repo in the mayor_finance_reports subdirectory.

I put all seven candidate files – and only those files – in that directory. So I know that I wanted to import the entire directory.

As covered in Chapter 7, base R's `list.files()` – or `dir()` – will list all files in a directory. `rio::import()` imports a *single* file into R, and purrr's map_df() function can help us run import() on a list of file names.

In Chapter 7, I showed you code to change the working directory to where the data files were stored. I often like changing to the subdirectory with the data, in case I want to save updated versions of the data in the same directory. However, changing working directories in R Markdown can be tricky. Another option is to stay in the project's parent directory, but ask list.files() to list files in a subdirectory *and return the file name including the full path to those files.* You can do that with list.files("subdirectory", full.names = TRUE):

```
contributions <- map_df(list.files("mayor_finance_reports",
                        full.names = TRUE), rio::import)
```

That code says "List each file from the mayor_finance_reports subdirectory, return the file names with full path to each file, run rio::import() on those complete file names, and save results to a data frame called contributions."

Do you find that including list.files() within map_df() makes the code difficult to read? If so, you can start with a line of code to create a separate variable holding all the file names, and then run map_df() on the new vector of file names:

```
mydatafiles <- list.files("mayor_finance_reports",
                 full.names = TRUE)
contributions <- map_df(mydatafiles, rio::import)
```

As always, there's a balance between tight, elegant code and code you're more likely to understand if you look at it next month. Unless you've got a massive data set where adding an extra variable will take significant time and memory, the only "right" answer is what works best for you.

Next, I wanted to keep only those contributions that were from Framingham, which I did using dplyr's filter() function. I could have added a new line of code: `contributions <- filter(contributions, City == "Framingham")`. Instead, I'm using using a `%>%` pipe workflow:

```
contributions <- map_df(list.files("mayor_finance_reports",
                        full.names = TRUE), rio::import) %>%
  filter(City == "Framingham")
```

For this particular analysis, I decided to eliminate Framingham addresses that were just post office boxes, since it's not clear someone with a Framingham post office box is actually a resident. I added code to my filter that said "keep only those records where the Address field, when converted to all lower case, doesn't contain "box".

These are important R functions to know when trying to create a filter like this:

- `!` means "NOT".
- `str_detect` is a function from the stringr package that checks whether a string contains a certain pattern – the syntax is `str_detect(mystring, mypattern)`.
- `tolower()` is probably fairly intuitive – it changes the character string to all lower case.

With all these skills in mind, here's the code for "keep records where the Address field, converted to lower case, doesn't contain 'box' ":

`!str_detect(tolower(Address), "box")`.

Updated code so far:

```
contributions <- map_df(list.files("mayor_finance_reports",
                 full.names = TRUE), rio::import) %>%
  filter(City == "Framingham", !str_detect(tolower(Address), "box"))
```

Next, I wanted to screen out duplicates. To eliminate duplicate *rows* in a data frame, I'd use base R's `unique()` function: `unique(contributions)`. However, checking for *fully* duplicated rows wouldn't help here, unless I'm looking for times when the same person gave the same amount on the same day twice (contribution date is one of the fields).

Instead, I'd like to remove duplicate rows based on just a few columns: In this case Contributor and Address. This may not be exact – it's possible that someone's name will be entered slightly differently in two different records; or, two different people in one apartment building could have the same name, if apartment numbers aren't included in an address. For my purposes in this analysis, checking name and address was close enough. However, if you are working on a project such as trying to match registered voters with a Social Security deaths database or comparing school bus drivers with drunk-driving arrests, more research would be needed to make sure you definitely have the same person both times.

dplyr's `distinct()` function will remove data frame rows based on duplication in certain columns. If you want to keep all the other variables in a data frame and not just the non-repetitive ones, distinct() needs the additional `.keep_all = TRUE` argument. Otherwise, it will return a data frame with only the columns you want to ensure haven't been duplicated.

Here, then, is the code version so far:

```
contributions <- map_df(list.files("mayor_finance_reports",
                 full.names = TRUE), rio::import) %>%
  filter(City == "Framingham", !str_detect(tolower(Address), "box")) %>%
  distinct(Contributor, Address, .keep_all = TRUE)
```

Finally, I'd like to count the number of rows in each candidate's group. I could do that in dplyr, but janitor's tabyl() function makes it even easier. The full code:

```
contributions <- map_df(list.files("mayor_finance_reports",
                 full.names = TRUE), rio::import) %>%
  filter(City == "Framingham", !str_detect(tolower(Address), "box")) %>%
  distinct(Contributor, Address, .keep_all = TRUE) %>%
  tabyl(Recipient, sort = TRUE)
contributions
```

```
##                  Recipient   n     percent
##       Horrigan, Joshua Paul  12 0.035820896
##  Neves-Grigg, Sr., Benjaman   4 0.011940299
##                Sen, Dhruba   3 0.008955224
##            Sousa, Priscila  10 0.029850746
##       Spicer, Dr. Yvonne M. 173 0.516417910
##          Stefanini, John A. 113 0.337313433
##             Tilden, Mark S.  20 0.059701493
```

If you'd like, you can add another line, `mutate(percent = round(percent * 100, 1))` to change the percent columns from a format like `0.50656168` to `50.7`. For now, though, I'll leave it as is, except for deleting the n column (I don't need the raw number of contributors) and renaming the columns to something more intutive.

dplyr's select() function can also *rename* columns in addition to *choosing to keep them*, with the syntax `select(mydataframe, mynewname = myoldcolname)`:

```
contributions %<>%
  select(Candidate = Recipient, Pct_Local_Contributors = percent)
```

If my theory was correct, Dr. Spicer – somewhat of a political novice who was vice president at Boston's Museum of Science – should be favored over Rep. John Stefanini, a well-known figure in town as a former state representative and member of the Board of Selectmen.

Who won Framingham's preliminary mayoral race? Election results are in the *election_framingham_mayor_2017_09.csv* file in the data subdirectory. I chose to use the readr package's read_csv() function so I can specify that the first row includes column names and isn't part of the data. I do this by adding the `col_names = TRUE` argument to read_csv(). If using rio::import(), the syntax would be

```
rio::import("data/election_framingham_mayor_2017_09.csv", header = TRUE)
```

That's because import() uses the data.table package's fread() function, not read_csv(). You can find out more about fread() arguments in the help file, running `?fread`.

For those who'd like to use base R, you can achieve something similar with:

```
results <- read.csv("data/election_framingham_mayor_2017_09.csv",
                     header = TRUE, stringsAsFactors = FALSE)
```

(stringsAsFactors = FALSE so that candidates' names come in as character strings, not factors).

Here's the code for readr::read_csv(), followed by a look at the column names with colnames():

```
results <- readr::read_csv("data/election_framingham_mayor_2017_09.csv", col_names = TRUE)
colnames(results)
```

```
##  [1] "Candidate" "1"          "2"          "3"          "4"
##  [6] "5"         "6"          "7"          "8"          "9"
## [11] "10"        "11"         "12"         "13"         "14"
## [16] "15"        "16"         "17"         "18"         "Totals"
```

Since I don't need results by precinct, I selected just the Candidate and Totals columns and then looked at the data:

```
results <- select(results, Candidate, Totals)
results
```

```
## # A tibble: 9 x 2
##   Candidate                 Totals
##   <chr>                      <int>
## 1 Blanks                        56
## 2 Joshua Paul Horrigan         545
## 3 John A. Stefanini           3184
## 4 Dhruba P. Sen                101
## 5 Mark S. Tilden               439
## 6 Yvonne M. Spicer            5967
## 7 Benjaman A. Neves-Grigg,     134
## 8 Priscila Sousa               538
## 9 Write-Ins                     42
```

I'd like to get percentages here. However, I can't use tabyl(), since this data frame already tallied up totals by candidate (if you remember, tabyl() counted rows by group). Instead, janitor's adorn_percentages() function will calculate percentages in a data frame, allowing you to choose whether the denominator for dividing each item should be a sum by `row` or `col`.

If you'd like the results to look like conventional percents – multiplied by a hundred and rounded with the percent sign included – add the adorn_pct_formatting() function. However, this will turn the percents into character strings because of the percent sign, which R doesn't recognize as part of a number. So only use that formatting if you don't need the data as numbers in your data frame.

The syntax for calculating percents that are non-rounded fractions: `adorn_percentages(mydf, denominator = "col")` for calculating by column. adorn_percentages(mydf) defaults by row.

```
results <- results %>%
  filter(!(Candidate %in% c("Blanks", "Write-Ins")) ) %>%
  adorn_percentages(denominator = "col")
results
```

```
##                 Candidate      Totals
##       Joshua Paul Horrigan 0.049963330
##          John A. Stefanini 0.291895856
##              Dhruba P. Sen 0.009259259
##             Mark S. Tilden 0.040245691
##           Yvonne M. Spicer 0.547029703
## Benjaman A. Neves-Grigg, 0.012284562
##            Priscila Sousa 0.049321599
```

Next, I'd like to keep the Candidate column as is, and change the name of the Totals column. dplyr's `rename()` function comes in handy – it renames columns you specify but otherwise keeps all the other data frame columns unchanged.

```
results <- results %>%
  rename(Pct_Vote = Totals)
```

I'd like to merge these two data frames into one, so I can compare the percent of local contributors with each candidate's percent of the actual vote. However, there's a real-world messy data problem in these data sets: Candidate names in the contributions data frame are in the format *Spicer, Dr. Yvonne M.* while results use the format *Yvonne M. Spicer.* How to fix it so they're the same?

17.5 Standardizing multiple versions of the same name

In the real world, with just seven candidates, it might be easiest just to cut and paste seven values from one Excel spreadsheet to another. But that's not very reproducible ... and it does introduce the possibility of error that an examination of your code wouldn't show.

Instead, knowing that no two candidates share the same last name, I'll transform each data frame's Candidate column to hold just their last names. Let's think about how to do this.

In the contributions data frame, the last name is at the *start* of each candidate name, up until the first comma. In the election results data frame, the last name is at *end* of each candidate name, from the final space until the end (with one name needing a comma removed). Let's begin with the contributions data.

tidyr's separate() function will split a data frame column into multiple columns based on a delimiter of your choice. The syntax is `separate(my_df, my_col_name, my_new_col_names, my_delimiter, my_number_of_new_columns)`. If I only create two columns separated by commas, I'll lose part of the candidate name "Neves-Grigg, Sr., Benjamin". Since I only need the last names, though, it won't matter that some of the first name will be dropped. In fact, I'll eliminate the FirstName column after creating it with `select(-FirstName)`. Here's the code, followed by a look at the first three rows:

```
contributions_split <- tidyr::separate(contributions, Candidate,
                         c("LastName", "FirstName"), ", ", 2) %>%
  select(-FirstName)
```

```
## Warning: Expected 2 pieces. Additional pieces discarded in 1 rows [2].
```

```
head(contributions_split, 3)
```

```
##      LastName Pct_Local_Contributors
##      Horrigan            0.035820896
##   Neves-Grigg            0.011940299
##           Sen            0.008955224
```

Note we get a warning that some data was lost in row 2 during the split (data which we know we don't need).

The results data frame is a bit more complicated, because one candidate has no middle name and thus only one space before her last name, while the others have first and middle names, and so two spaces before their last names.

Probably the easiest way to deal with this based on skills we've already learned is to split by space anyway. Then, if the LastName column is empty, fill it with the MiddleName column. That will handle the case of the candidate without a middle name.

Here's how to split the Candidate column into three different columns:

```
results_split <- tidyr::separate(results, Candidate, c("FirstName", "MiddleName",
"LastName"), " ")
```

```
## Warning: Expected 3 pieces. Missing pieces filled with `NA` in 1 rows [7].
```

```
tail(results_split, 3)
```

```
## # A tibble: 3 x 4
##   FirstName MiddleName LastName     Pct_Vote
##   <chr>     <chr>      <chr>           <dbl>
## 1 Yvonne    M.         Spicer         0.547
## 2 Benjaman  A.         Neves-Grigg,   0.0123
## 3 Priscila  Sousa      <NA>           0.0493
```

And here is code for populating the LastName with MiddleName if it's empty, deleting the first- and middle-name columns, and stripping the comma off the end of Neves-Grigg,:

```
results_split %<>%
  mutate(
    LastName = ifelse(is.na(LastName), MiddleName, LastName),
    LastName = str_replace(LastName, ",", "")
  ) %>%
  select(-FirstName, -MiddleName)
```

What does that code do? The first line sets the source and destination data frames to both be results_split. The first line after mutate says "If the value of LastName is NA, then the value should be what's in the MiddleName column. If not, set the value to the existing LastName value".

The second line replaces any comma in LastName with nothing. Finally, the select() function removes two columns, FirstName and MiddleName.

17.5.1 For advanced (or fearless) readers: regular expressions

In the real world, I'd use what are called regular expressions to extract last names from these character strings.

Regular expressions are not R-specific. They're implemented in almost every modern programming language – and in many text editors, such as Notepad++ for Windows and TextWrangler for Mac. Teaching regular expressions is *far* beyond the scope of this book, but there are a few resources at the end of this chapter if you'd like to learn more.

To briefly explain: Regular expressions let you look for *patterns*, not just specific characters. Instead of searching for a space character (or a specific word such as "Spicer"), regular expressions can search for things such as "any four letters in a row" or "three digits followed by a hyphen". Or, in this case "all the characters that follow the last space in a character string".

Like a lot of things in R, regular expressions are slightly different than in many other languages. For example, the first match of a pattern within parentheses is referred to as `\1` or `$1` in most languages. But in R, you need `\\1`.

Here's the regular expression to find "all the characters that follow the last space in a character string until the end of the string, deleting a comma if there's a comma at the end": `.*\\s(.*?)\\,?$` .

To deconstruct that: A dot stands for 1 character of any kind. The asterisk that follows means "0 or more of whatever was before". So, `.*` means "0 or more of any character." The `?` afterwards means "the earliest possible match" because by default, regular expressions will include as much as possible when looking for a match, something known as "greedy" matching.

`\\s` means "a space". The parentheses around `.*?` signify that this is the first (and in this case only) part of the match I want to extract. Since it's the first, R will refer to it as `\\1` .

Finally, `\\,` means a comma (the `\\` is needed to "escape" the comma, and tell R that I want an actual comma, since commas have special meanings in regular expressions). And `$` signifies "end of the string".

This is how I'd use a regular expression to extract last names from the results$Candidate column:

```
results_regexp <- results %>%
  mutate(
    LastName = str_replace_all(Candidate, ".*\\s(.*?)\\,?$", "\\1")
  )
tail(results_regexp, 3)
```

```
## # A tibble: 3 x 3
##   Candidate                  Pct_Vote LastName
##   <chr>                         <dbl> <chr>
## 1 Yvonne M. Spicer             0.547  Spicer
## 2 Benjaman A. Neves-Grigg,     0.0123 Neves-Grigg
## 3 Priscila Sousa               0.0493 Sousa
```

17.6 Making 2 data frames 1

Now that the two data frames have a common column in the same format, we can merge them into one. We saw how to do this in Chapter 8 with dplyr's left_join() or base R's merge().

Reminder that the merge() syntax is `merge(df1, df2, by.x = "col1", by.y = "col2", all.x = TRUE, all.y = TRUE)` where by.x is the name of the merge column in the first data frame, by.y is the column in the second data frame, and all.x and all.y decide whether rows without a match in the other data frame should be included in the result.

Running `?join` in the console will give you a rundown of dplyr options, including `left_join()` for all rows in the first data frame, `inner_join()` to return rows from the first data frame where there are matches in the second, and full_join() returning all rows and columns in both data frames. There should only be 7 rows after the join; I'll use a full_join() to make sure:

```
mayordata <- full_join(contributions_split, results_split, by = "LastName")
str(mayordata)
```

```
## 'data.frame':    7 obs. of  3 variables:
##  $ LastName              : chr  "Horrigan" "Neves-Grigg" "Sen" "Sousa" ...
```

```
## $ Pct_Local_Contributors: num 0.03582 0.01194 0.00896 0.02985 0.51642 ...
## $ Pct_Vote              : num 0.04996 0.01228 0.00926 0.04932 0.54703 ...
```

```
head(mayordata)
```

```
##      LastName Pct_Local_Contributors    Pct_Vote
## 1    Horrigan            0.035820896 0.049963330
## 2 Neves-Grigg            0.011940299 0.012284562
## 3         Sen            0.008955224 0.009259259
## 4       Sousa            0.029850746 0.049321599
## 5      Spicer            0.516417910 0.547029703
## 6   Stefanini            0.337313433 0.291895856
```

If you don't want to remember which join is which and would rather use merge() and arguments all.x, all.y, by.x, and by.y, the format would be

```
mayordata <- merge(contributions_split, results_split, all.x = TRUE, all.y = TRUE,
                   by.x = "LastName", by.y = "LastName")
```

17.7 Analyzing and graphing the results

At long last we can analyze and visualize the results. What's the correlation between percent local contributors and percent of the final vote? Base R's `cor()` function creates a *correlation matrix* of each variable vs. all others. It takes a matrix or data frame with all numerical values. In this case, we've only got two variables. We can run it anyway on the mayordata data frame, as long as we remove the non-numerical LastName column. We can do that easily with base R, removing the first column `mayordata[,-1]` or, more elegantly, with dplyr's select:

```
select(mayordata, -LastName) %>%
  cor()
```

```
##                        Pct_Local_Contributors  Pct_Vote
## Pct_Local_Contributors              1.0000000 0.9920022
## Pct_Vote                            0.9920022 1.0000000
```

Whoa, that's a pretty powerful relation. Perfect correlation, as you see in the columns where each variable is compared with itself, would be 1.0, while absolutely no relationship at all would be 0. (Perfect negative correlation where one value goes down in lockstep as another goes up would be -1). The correlation here between percent local contributors and percent of the preliminary election vote is more than 0.99.

17.8 Visualizing results

You probably don't need a graph to detect a relationship in data where the correlation is more than 0.99. However, it can help to see which candidates under- or over-performed on Election Day compared with their number of local contributors.

For that, you can use a ggplot2 scatter plot, adding geom_smooth(method='lm') for a linear regression line (Figure 17.1). That generates a straight line with the least amout of space between the line and each point (more specifically, the smallest sum of the squared distance between each point and the line). I'll turn off the confidence level shading with `se = FALSE`.

```
ggplot(mayordata, aes(Pct_Local_Contributors, Pct_Vote)) +
  geom_point() +
 geom_smooth(method='lm', se = FALSE)
```

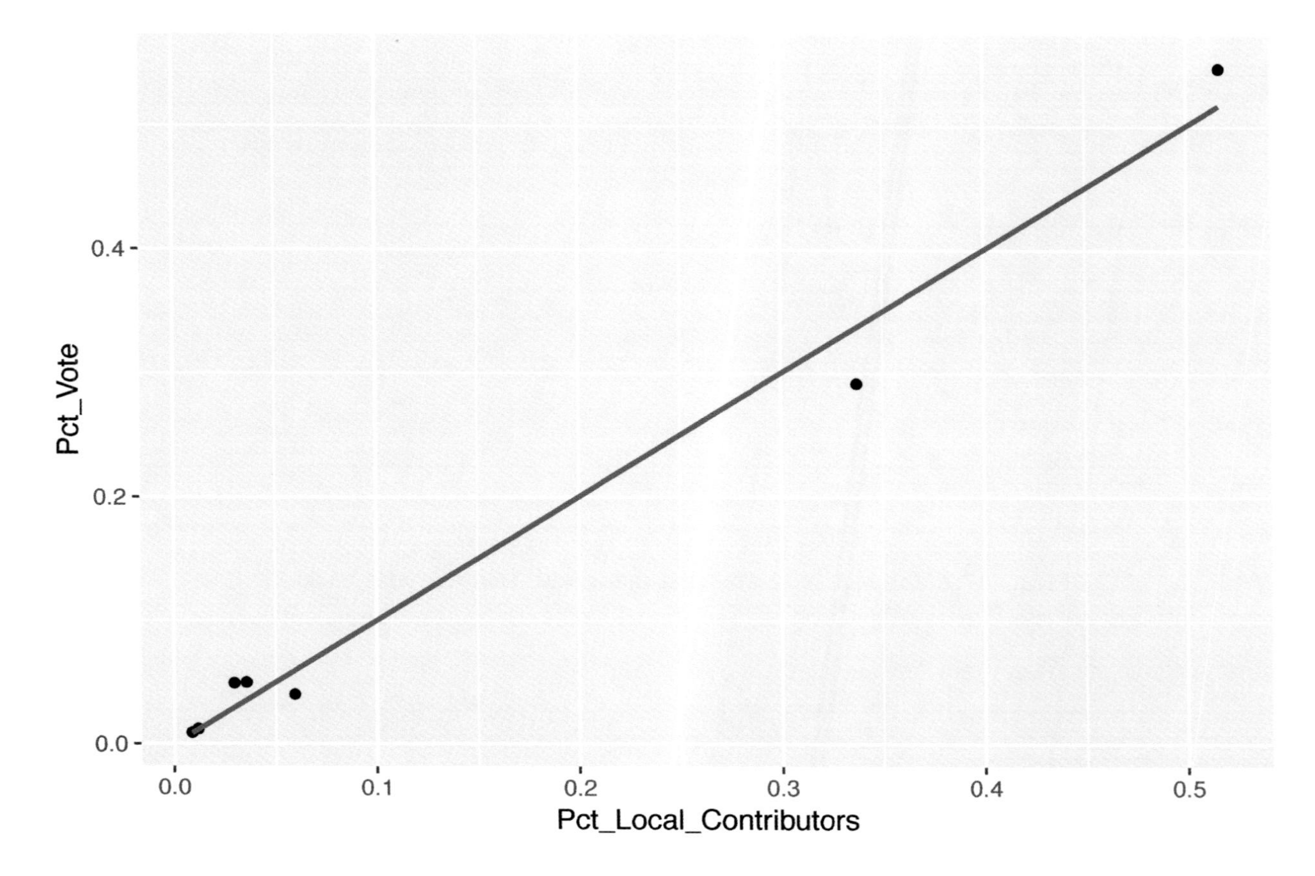

Figure 17.1: Scatter plot of percent local contributions on the x axis and percent vote on the y axis, with a linear regression line added.

It would be useful to make this interactive, so that rolling over points displays the underlying data.

The taucharts package is one of my favorite choices for generating a JavaScript interactive scatterplot in R. The tau_tooltip() function adds a tooltip that includes *all* the variables in the data frame (ggplotly, discussed in Chapter 9, would only show the x and y numerical values by default). And the tau_trendline() function creates a dropdown menu offering different types of trend lines (see Figure 17.2).

```
library(taucharts)
tauchart(mayordata) %>%
  tau_point("Pct_Local_Contributors", "Pct_Vote") %>%
  tau_tooltip() %>%  # includes all variables in mydf
  tau_trendline()
```

17.9 Consider R Markdown

This chapter's entire analysis could be contained in an R Markdown document. The advantage of R Markdown is that you can write your story, blog post, press release, or narrative, and incorporate R code within the same document. Figure 17.3 looks at part of the R Markdown file I used on my blog the day after the preliminary election:

I used an R Markdown document to write this chapter – and the entire book. I've included a simple R Markdown file called packages.Rmd that generates an HTML file with an interactive table of packages in this

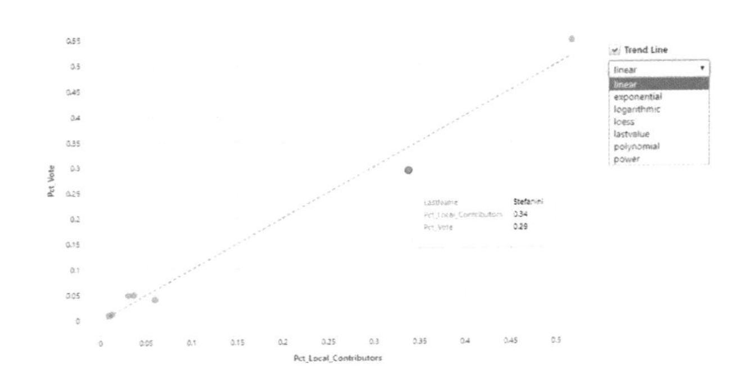

Figure 17.2: Interactive scatter plot made with the taucharts package.

```
```{r}
thecor <- round(cor(votes2$PercentFramDonors, votes2$PctPreliminaryVote),3)
```

In fact, there's a statistical correlation of `r thecor` -- and the highest
possible number is 1.

This doesn't mean I believe that the number of donors was the factor
responsible for voting results. As the saying goes, "correlation is not
causation." What this _does_ mean is that _in this particular instance,_
knowing the percent of local donors a candidate had could help you predict
that candidate's percent of vote tonight. I'm not ready to conclude there
will be a similar correlation in the November 7 general election, though.

Nevertheless, the correlation in this preliminary election was pretty
striking:

```{r table, echo=FALSE, results='asis', warning=FALSE, message=FALSE}
DT::datatable(votes[,-2], rownames = FALSE, options = list(dom = 't'))
```


<center>

```{r scatterplot, echo=FALSE, results='asis', warning=FALSE, message=FALSE}

plotly::ggplotly(
  ggplot(votes2, aes(x=PercentFramDonors, y=PctPreliminaryVote, tooltip =
Recipient)) +
    #  geom_point(size=2.5) +
    geom_point() +
    theme_minimal() +
    xlab("Percent Framingham Donors") +
    ylab("Percent Preliminary Election Vote") +
    ggtitle("Percent Framingham Donors vs Percent Vote"),
  width = 440, height = 425
)

```

</center>
```

Figure 17.3: Portion of an R Markdown document for analyzing election results vs. local campaign contributions.

book. You can open it, see how it's formatted, and click the Knit button to generate an HTML page.

Next (and last) up: A few more R tips and tools, as well as several inspiring data journalism articles that used R for analysis.

17.10 Additional resources

Gloria Lin and Jenny Bryan posted a **regular expressions tutorial** as part of a University of British Columbia Stats 545 course. It's available online at http://stat545.com/block022_regular-expression.html.

The **RegExplain RStudio addin** features a regex cheat sheet reference and an interactive regular-expression builder that shows you text matches as you work. There's a good explainer in the readme file at https://github.com/gadenbuie/regexplain/#readme. (For general regex-building without R-specific quirks, regex101.com is a useful site.)

Chapter 18

Additional resources

I hope you feel like you've gotten a solid start in using R for analyzing data and effectively communicating the results of that analysis! Of course, like with any technical skill, there's always more to learn.

Here are a few additional tips, functions, tutorials, and inspiring R-based journalism examples to help you continue your R journey.

18.1 More functions, packages and tools worth a look

datapasta - This package makes it easy to copy data into your clipboard and then generate R code to create a vector or data frame from that information. (Aside: What's not to like about a GitHub repo that starts off "On top of spaghetti, all covered in cheese...") See more from author Miles McBain at https://github.com/MilesMcBain/datapasta.

The RStudio addin **Bare Combine** lets you select a string of comma-separated characters such as `New York, Boston, Chicago` and turn them into a properly quoted R vector `c("New York", "Boston", "Chicago")`. This requires RStudio. " 'Bare Combine' is going to change my life," New York Times Ben Casselman tweeted after first reading about it (in my Computerworld roundup of useful R functions). Bare Combine author Bob Rudis explains how it works as part of his hrbraddins package at https://github.com/hrbrmstr/hrbraddins.

To use an RStudio addin after installing it, you can look for it under the Addins dropdown (such as in Figure 18.1) and select it from the menu. Even easier, though, is to **create an RStudio keyboard shortcut.** As mentioned briefly in Chapter 2, you can create your own keyboard shortcuts within RStudio by heading to Tools > Modify Keyboard Shortcuts. You'll see a rather lengthy list of possible actions to shortcut; scroll down until you get to addins. Once you find the one you want, double click in the shortcut area and choose the key combination you'd like.

In one of the most meta packages ever, Dean Attali created **an addins package that is itself an addin**, allowing you to browse lots of available addins you might want to install. You can install it with `install.packages("addinslist")` or `pacman::p_install(addinslist)`.

It seems that almost every other month brings another compelling R package that generates interactive JavaScript-based visualizations. One intriguing (relative) newcomer is **echarts4r**, an R wrapper around the powerful and flexible echarts JavaScript library. Creating basic graphs and tooltips is easy, although generating a scatter plot with a tooltip that includes more than two data points takes a little more code than taucharts. Find out more at http://echarts4r.john-coene.com. Code for a scatterplot from the mayoral data in the previous chapter is in the echarts4r.R file in the repo.

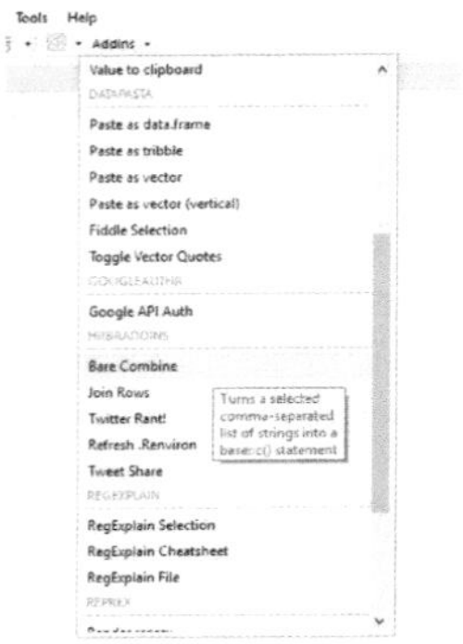

Figure 18.1: RStudio addins menu

18.2 Stories done with R

"Huge increase in arrests of homeless in L.A. — but mostly for minor offenses", Los Angeles Times, February 4, 2017. Arrest data was analyzed with R, and the code is available at the LA Times' Data Desk GitHub repository: https://github.com/datadesk/homeless-arrests-analysis.

"How the Cook County Assessor Failed Taxpayers", ProPublica, Dec. 7, 2017. Investigation of Cook County's "error-ridden commercial and industrial assessments". Story: https://features.propublica.org/the-tax-divide/cook-county-commercial-and-industrial-property-tax-assessments/. Code: https://github.com/propublica/propertyassessments.

"Roger Federer: 20 Years, 20 Titles", SRF (Swiss Radio & TV), January 28, 2018. Detailed look at Federer's career. Story in English: https://www.srf.ch/static/srf-data/data/2018/federer/en.html#/en/. Methodology: https://srfdata.github.io/2018-01-roger-federer/

18.3 Tutorials

18.3.1 General

R, RStudio, and the tidyverse for data analysis - Well documented, easy-to-follow beginner tutorial presented at the 2018 Investigative Reporters and Editors' Computer Assisted Reporting conference. By Peter Aldhous, BuzzFeed News. http://paldhous.github.io/NICAR/2018/r-analysis.html

File organization best practices: How to set up a reproducible workflow in R. Also presented at the 2018 CAR conference. By Andrew Tran, Washington Post. https://andrewbtran.github.io/NICAR/2018/workflow/docs/01-workflow_intro.html.

R for Data Science - This book by Hadley Wickham and Garrett Grolemund will give you a good grounding in R fundamentals as you learn the basics of analyzing data in R. It's available online for free at http://r4ds.had.co.nz/

A community launched around the book led by Jesse Maegan in 2017 and exploded in popularity the following year. You can find out about the Slack channel at https://medium.com/@kierisi/r4ds-the-next-iteration-d51e0a1b0b82. There's also a Twitter account @R4DScommunity.

Wickham recorded an informal, 20-minute video to show his start-to-finish workflow for initially exploring a data set, in this case U.S. building permits. The companion GitHub repo is at https://github.com/hadley/building-permits, which includes a link to the video.

I've been creating a weekly screencast for InfoWorld, Do More With R, that aims to show interesting tips in 5 minutes or less. Topics have included using dplyr's case_when() function, testing code with testthat, and creating dashboards with the flexdashboard package. https://bit.ly/morewithR.

18.3.2 Misc

If there's one task that almost everyone needs to do in R at one time or another, it's reorder bars on a graph. This tutorial tells you everything you need to know about using the tidyverse forcats package for reordering ggplot2 bar charts: https://github.com/jtr13/codehelp/blob/master/R/reorder.md

18.3.3 Shiny Web framework

It's possible to create fully featured Web applications using R, thanks to the Shiny Web framework. There's a lot of information about Shiny, including a number of tutorials, at https://shiny.rstudio.com.

In addition, DataCamp is offering a free online class, including access to their R cloud platform, to learn interactive Web programming with Shiny. You can search for "Building Web Applications in R with Shiny" at datacamp.com, or head directly to https://www.datacamp.com/courses/building-web-applications-in-r-with-shiny.

18.4 Social media, communities, and Web resources

I'm the administrator for the R for Journalists Google Group. It's a private group just to limit spam, but anyone with an interest in journalism and R is welcome. Apply here: https://groups.google.com/d/forum/rjournos.

The RStudio Community is a place to ask questions about RStudio's products and packages specifically but also R in general. Find it at https://community.rstudio.com/

The #rstats hashtag on Twitter is a good place to find out about R news and developments. People post about new packages, great tutorials, and more. The Google Plus R group is another good resource.

If you've found an issue with code or other content in the book, I'm tracking them at the book's GitHub repository. You don't need to know git in order to open an issue; all you need is a browser and free GitHub account. Post an issue at https://github.com/smach/R4JournalismBook/issues. I will be posting any post-publication updates or corrections to the repo's book_updates_and_corrections.Rmd file.

Good luck with your continued R journey! I'd love to hear from you on Twitter @sharon000 or by email at r4journos@machlis.com.

Appendix A Online: How do I . . .

The book's GitHub repository includes a searchable version of this appendix of key tasks covered in the book, including sample code for some tasks as well as the chapter listings. Look for the file HowDoI.html if you downloaded the repo to your local system, or head to https://smach.github.io/R4JournalismBook/HowDoI.html.

Tasks and chapter(s) where they were covered

Add a CSS stylesheet to an R Markdown document's YAML header, 15

Add a linear regression line to a ggplot2 visualization, 12

Add a pre-built theme to an R Markdown document's YAML header, 15

Add a total column to a data frame, 14

Add a total row to a data frame, 14

Add address search to a leaflet map, 11

Add comments to code, 6

Add data to a geospatial object, 11

Add or subtract days, weeks, or years from a date, 13

Add or subtract months from a date, 10, 13

Add some randomness to a ggplot2 scatter plot so points don't cover up other points, 9

Apply a function to a data frame by group, 8

Apply a function to all items in a data frame by column or row, 15

Apply a function to all items in a vector or list and return a character vector, 7

Apply a function to all items in a vector or list and return a data frame, 7

Apply a function to all items in a vector or list and return a numeric vector, 7

Apply a function to all items in a vector or list when you don't want to store the output, 15

Apply a function to two vectors in parallel, 16

Assign a character string to a variable, 2

Assign a numerical value to a variable, 2

Calculate candidate vote percent by row, 12

Calculate correlations, 12

Categorize values into intervals, 10

Change or set the projections of an sf geospatial object, 11

Appendix B Online: Functions

A searchable table of functions covered in the book (and chapters where they appeared) is available in the functions.html file or at https://smach.github.io/R4JournalismBook/functions.html.

Function, Package, Chapter

%>%, magrittr and dplyr, 5

%<>%, magrittr, 12

%in%, base R, 7

?, base R, 2

??, base R, 2

<-, base R, 2

addMarkers(), leaflet, 11

addSearchGoogle(), leaflet.extras, 11

addSearchOSM(), leaflet.extras, 11

addTiles(), leaflet, 11

adorn_totals(mydf), janitor, 14

anti_join(df1, df2, by = c("df1col" = "df2col")), dplyr, 8

any(myvec == "mycondition"), base R, 13

append_data(shape_object, data_object, key.shp = colname, key.data = colname), tmaptools, 11

apply(mydata, margin, myfunction), base R, 15

arrange(mydf, col), dplyr, 5

as.character(mycol), base R, 5

as.Date("yyyy-mm-dd"), base R, 13

as.numeric(mycol), base R, 5

barplot(myvec), base R, 6

basename("http://myurl.com/filename.html"), base R, 16

bind_rows(), dplyr, 7

boxplot(myvec), base R, 6

c(), base R, 2

case_when(), dplyr, 10

Appendix C Online: Packages

A searchable *table of packages covered in the book (and chapters where they first appeared) is available in the packages.html file or online at https://smach.github.io/R4JournalismBook/packages.html.*

Package, Chapter

cancensus, 4

corrplot, 12

corrr, 12

data.table, 4

dplyr, 5, 6, 7, 8, 10, 13

DT, 12

dygraphs, 3

echarts4r, 18

esquisse, 6

eu.us.opendata, 4

feather, 4

fivethirtyeight, 4, 12

forcats, 6

fst, 4, 15

geofacet, 9

gganimate, 9

ggiraph, 9

ggmap, 9, 11

ggparliament, 12

ggplot2, 6, 8, 9, 12

ggrepel, 9

glue, 11

googlesheets, 4

here, 7, 17

highcharter, 12

Index